“A JOURNEY TH
VIRTUAL UN

— *Wall Street Journal*

PRAISE FOR

INFINITE REALITY

“An exhilarating book. . . . Blascovich and Bailenson are ideally situated to write this guide to the new world. . . . *Infinite Reality* is a must-read for anyone who wants to prepare for the coming revolution.”

— *Los Angeles Times*

“Read this book if you want to understand the future. Virtual reality is changing our world, both because it's one of the design ideals that digital devices are evolving toward and because it provides a mythic undercurrent to the culture of technology. And yet before this book, too little had been written to explain what VR is and how it opens a new window on understanding human culture and cognition. The authors are bright lights in the emerging discipline of using VR to understand people.”

— Jaron Lanier, author of *You Are Not a Gadget*

“Brilliant, farsighted, and fascinating, *Infinite Reality* is a must-read exploration of the coming human and social transformations driven by digital immersive virtual reality technology. Jim Blascovich and Jeremy Bailenson are leading minds in this evolving and breathtaking world, and *Infinite Reality* is an essential guide to our futures.”

— Philip Zimbardo, bestselling author of *The Lucifer Effect* and professor emeritus of psychology at Stanford University

“Have you ever become immersed in virtual reality? It is an overpowering experience, and in *Infinite Reality* Jeremy Bailenson and Jim Blascovich capture what it is like in words. Their crystal-clear descriptions send chills down the spine as we come to realize that our biological reality is probably virtual as well. We will soon be at one with these technologies, as we will seek from them what we seek from our natural world and kin.”

— Michael S. Gazzaniga, author of *Human: The Science Behind What Makes Your Brain Unique* and director of the SAGE Center for the Study of the Mind at the University of California–Santa Barbara

"Enough with speculation about our digital future. *Infinite Reality* is the straight dope on what is and isn't happening to us right now, from Jeremy Bailenson and Jim Blascovich, two of the only scientists working on the boundaries between real life and its virtual extensions."
— Douglas Rushkoff, author of *Program or Be Programmed*

"A page-turner that immerses the reader into the world of virtual reality, Jim Blascovich and Jeremy Bailenson's *Infinite Reality* compellingly highlights the fine line between science and science fiction."
— Gary Small, MD, author of *iBrain: Surviving the Technological Alteration of the Modern Mind* and professor of psychiatry at UCLA's Geffen School of Medicine

"Two of virtual reality's most prominent researchers, Jim Blascovich and Jeremy Bailenson detail the current research—primarily their own—and pose some fascinating questions, which have surprising and important answers. . . . *Infinite Reality* illustrates groundbreaking research with vivid imagery and inspiring analysis." — *New Scientist*

"In their book *Infinite Reality*, pioneers Jim Blascovich and Jeremy Bailenson provide compelling evidence that virtual reality is not only the next step of our cultural evolution, but it is also as real and possible as the iPad or Kindle you are reading this on. From the first page, Professors Blascovich and Bailenson prove their worth as brilliant academics and educators and take the reader through a crash course on the past, present, and future of virtual reality. . . . The philosophical and ethical issues this book tackles are beyond the scope of almost any discussion you will find elsewhere. . . . *Infinite Reality* represents a clear, insightful, and comprehensive view on a form of technological advancement that has already begun to take shape. Are you ready?"
— *Cyberpsychology*

"Illuminating. . . . A sweeping presentation of virtual reality's ability to create new and multiform experiences and perspectives—likely to beguile more than a few skeptics."
— *Kirkus Reviews*

"The word *avatar* in the subtitle should be enough to get many readers to open the book, but there are many other reasons, too. . . . An imaginative and exciting look at the future that not just could be but, in many cases, already is." — *Booklist*

INFINITE REALITY

THE HIDDEN BLUEPRINT OF OUR VIRTUAL LIVES

JIM BLASCOVICH AND JEREMY BAILENSON

wm
WILLIAM MORROW
An Imprint of HarperCollins *Publishers*

To our families

Credits: page 12: courtesy of Daniel Simons; pages 14, 16, 78, 118, 177, 228: courtesy of Crystal Nwaneri; pages 17, 39, 43, 46, 66, 77, 85, 99, 111, 162, 203, 257: courtesy of Cody Karutz; page 29: courtesy of Joseph Nicéphore Niépce; pages 34, 214: courtesy of the U.S. Army; page 50: courtesy of Joseph Binney; pages 51, 258: courtesy of Julio Mojica; page 52: courtesy of Deonne Castaneda; page 56: courtesy of Felix Chang; page 58: courtesy of WorldViz LLC; pages 61, 104, 110: courtesy of Michelle Del Rosario; page 63: courtesy of Amanda Schwab; page 68: courtesy of Kip Williams; page 72: courtesy of Frank Goss; page 90: courtesy of Kim Swinth; page 97: courtesy of Ipke Wachsmuth; pages 129, 130, 237: courtesy of Cade McCall; page 149: courtesy of the Electronic Visualization Laboratory at the University of Illinois at Chicago and the University of Central Florida; page 150: Einstein head courtesy of David Hanson, Hanson Robotics, walking body courtesy of the Korea Advanced Institute of Science and Technology (KAIST); page 196: courtesy of Amanda Schwab and Crystal Nwaneri; page 201: courtesy of Andrew C. Beall; page 208: © Hunter Hoffman; page 210: courtesy of Albert "Skip" Rizzo; page 216: courtesy of Ron Artstein; page 223: © 2010 Learning Sites, Inc.

A hardcover edition of this book was published in 2011 by William Morrow, an imprint of HarperCollins Publishers.

FIRST WILLIAM MORROW PAPERBACK EDITION PUBLISHED 2012.

Designed by Jamie Lynn Kerner

Library of Congress Cataloging-in-Publication Data has been applied for.

ISBN 978-0-06-180951-4 (pbk.)

12 13 14 15 16 OV/RRD 10 9 8 7 6 5 4 3 2 1

CONTENTS

INTRODUCTION

"RIGHT NOW, WE'RE INSIDE A COMPUTER PROGRAM?"

With that monotone query, a very confused Neo, played by Keanu Reeves in the blockbuster film *The Matrix*, convinces hundreds of millions of viewers that virtual reality could be so real that people have no idea they are actually living in a simulation. Of course, *The Matrix* is just a movie, but brain science supports many of the ideas of the Wachowski brothers, who wrote, directed, and produced the film.

The brain often fails to differentiate between virtual experiences and real ones. The patterns of neurons that fire when one watches a three-dimensional digital re-creation of a supermodel, such as Giselle or Fabio, are very similar—if not identical—to those that fire in the actual presence of the models. Walking a tightrope over a chasm in virtual reality can be a terrifying ordeal even if the walker knows it's virtual rather than physical.

People interact via digital stimuli more and more. According to a recent study by the Kaiser Family Foundation, kids spend eight hours per day on average outside of the classroom using digital media. This translates to billions of hours per week. People interact with virtual representations in just about every facet of life—business transactions, learning, dating, entertainment, even sexual relationships. Online dating, which used to be somewhat stigmatizing, is now normative. Young adults consider their Facebook friends just as important as the people who live close enough to meet physically. In the world of online games and virtual worlds, millions of players spend over twenty hours each week "wearing" avatars, digital representations of themselves. Strikingly, the average age of these players is not fifteen but twenty-six. Household "console" video arenas, especially games, in which people control and occupy avatars, consume more hours per day for kids than movies and print media combined. To borrow a term from the new vernacular, virtual experiences are spreading virally.

Technological developments powering virtual worlds are accelerating, ensuring that virtual experiences will become more *immersive* by providing sensory information that makes people feel they are "inside" virtual worlds. In the United States, Nintendo's Wii, often coupled with a huge high-definition television, populates many living rooms. The players' physical actions are transformed into virtual body movements in the game. By the time you read this, Nintendo's Wii, Microsoft's Kinect, and Sony's PlayStation Move may incorporate 3-D displays. Virtual experiences are no longer embodied just by hunting and pecking on a keyboard or using a joystick: digital characters now move in tandem with players as they jump around, point guns, and swing racquets, golf clubs, and baseball bats.

Stereo, 3-D visual media technology—which not that long ago was only available to scientists and people using View-Masters—

promises to change the film, television, and game industry. Movie theaters entice audiences willing to pay a few extra dollars for 3-D glasses to watch blockbuster films. The game and television industry are promoting 3-D monitors to every household. The popular sports network ESPN even broadcasts in 3-D.

Although we aren't yet "jacking in" to the virtual world via a plug in the back of our head, as Neo did in *The Matrix*, digital media are providing more realistic experiences and not just for humans. Ten years ago, most household pets ignored television. Today, high-definition television transfixes, thrills, and sometimes enrages dogs and cats as they watch the fare on the Animal Planet network. They simply do not differentiate the digital image from reality.

This leads to an interesting proposition—the brain doesn't much care if an experience is real or virtual. In fact, many people prefer the digital aspects of their lives to physical ones. Imagine you never aged, could shed pounds of cellulite, or put on muscle mass at the touch of a button. Think about never having a bad-hair day, expressing an involuntary grimace, or getting caught staring. Think also about a world with no putrid smells but plenty of delightful ones, when it rains only when you are inside, and where global warming is actually just a myth. In this world, your great-grandfather is still around and can play catch with your six-year-old daughter. There is no dental drill or swine flu in this place.

But there are consequences to people occupying idealized digital worlds. This quandary is thematic in James Cameron's film *Avatar*, which took in more money than any prior film in United States history. In it, Jake Sully, a paraplegic soldier confined to a wheelchair, dons a virtual body of a member of another species, the Na'vi, on a distant planet. With avatar arms and legs, as well as a tail, he runs through jungles and swings through trees. He even falls in love.

On the one hand, *Avatar* depicts many wonderful aspects of virtual reality. In the natural world, physically disadvantaged people are denied many behaviors that most take for granted. In the virtual world, people can choose whether their avatars have fully functioning bodies, regardless of their physical condition. One of the most popular virtual worlds, *Second Life*, has a higher proportion of physically challenged users than the general population, allowing them to shed any stigmatization they experience in the physical world. Paraplegics can not only walk and run again, but actually can fly through the air or teleport themselves thousands of (virtual) miles in an instant.

On the other hand, Jake learns that wearing his Na'vi avatar has emotional consequences. He is a human being at the beginning of the movie, but as he spends more and more time wearing his giant blue alien avatar, he loses his humanity. By the end of the film, Jake's psychological bond with his avatar is so strong that he abandons his ties to the human race.

Avatar's fiction is supported by science: dozens of psychological experiments have shown that people change after spending even small amounts of time wearing an avatar. A taller avatar increases people's confidence, and this boost persists later in the physical world. Similarly, a more attractive avatar makes people act warm and social, an older avatar raises people's concern about saving money, and a physically fit avatar makes people exercise more.

Outside of scientific laboratories, avatars can be a matter of life or death. On the positive side, an avatar can be immortal. Consider the case of Orville Redenbacher, who is still the spokesperson for the popcorn company, even though he passed away years back. Using video footage from commercials starring Mr. Redenbacher, advertisers were able to construct a digital model that looks just like him and can be animated to perform any action imaginable. So the popular spokesperson is now "acting" in new advertisements from beyond the grave. There are commercial services today that will "immortalize" anyone who would like their avatars created and stored.

On the negative side, avatars can be sources of trauma. Consider the horrific case of a thirteen-year-old girl who committed suicide when she found out the "boy" with whom she interacted online wasn't who she thought he was. He was a fictional character created by others, who planned to hurt her feelings. She formed a strong attachment to the online persona. When she discovered he was fictional, she was devastated. In a less tragic but still disturbing event, in the early days of the Internet, there was a well-known rape case in cyberspace, in which one online user, via text, violated another in a virtual chat room. The victim, while physically unharmed, was traumatized.

Avatars also have the distinction of being completely anonymous but inherently "trackable." One can wear an avatar of any gender, age, race, species, or shape, and via the avatar, it is possible to meet others in virtual spaces without them having a clue about one's physical attributes and identity. On the other hand, any time people use the Internet, they leave a record behind (think "cookies" on Web browsers). Similarly, but in much greater detail, any time people enter a virtual space, they leave "digital footprints"—all the data the computer automatically collects: for example, speech, nonverbal behavior, and location. This footprint can be used (and, in fact, is being used) by military and other government agencies to detect identity. In essence, while one can hide behind an avatar of a different name, the footprint still can give him away.

In 1938, a carefully crafted radio broadcast caused millions of people to question their ability to differentiate the real from the virtual. Many of these listeners experienced emotions far worse than doubt and confusion—they were terrified. Orson Welles, via radio broadcast, presented a highly realistic, news-style depiction of an alien invasion in an adaptation of the novel *The War of the Worlds.* Though the program was intended as entertainment, those who had not heard the lead-in to the show thought the broadcast was an actual newscast. So many people panicked and fled in their cars that highways were flooded with traffic. Others aimed their rifles and shot at water towers that resembled spacecraft, or wrapped towels around their heads to protect themselves from potential alien mind-control. Even scientists were fooled. Several geologists rushed to the alleged scene in New Jersey to examine the fallen meteorites surrounding the alien craft. In sum, a well-crafted virtual story galvanized a large population.

The *War of the Worlds* calamity highlights why today's virtual revolution is particularly potent. In 1938, there was a clear distinction between media producers and media consumers. In order for *The War of the Worlds* to reach people's homes, corporate support was required. The show's producer, CBS, was one of the very few organizations that had access to airwaves. Because only a handful of program directors decided what types of stories would be broadcast, maintaining rational control over media content was possible—though not foolproof, as the broadcast's hysteria proved.

Contrast that with today's world, in which consumers are also media producers. Try to find a college student without an elaborately constructed Facebook profile. It won't be easy. Students constantly update photographs and diary entries for the world to see. Similarly, YouTube videos, produced by anyone with a Web connection and a digital camera, can receive worldwide attention just hours after being produced. The people who use the Web also shape the content of the Web. Sometimes those people become multimillionaires—for example, the creators of the game *Farm-*

Ville, a simple Facebook app that may have more farmers than the planet does.

We sit on the cusp of a new world fraught with astonishing possibility, potential, and peril as people shift from face-to-face to virtual interaction. If by "virtual" one means "online," then nearly a third of the world's population is doing so already. More than 300 million Web sites and numerous online applications, including e-mail, chat rooms, video conferencing, computer games, and social networking, keep over a quarter of the world's nearly 7 billion humans busy—in some cases, obsessively—interacting virtually. Users average three hours per day online. In countries like South Korea, the average is much higher. Digital interactions among people are becoming ubiquitous at work and play. The vice president of Digital Convergence at IBM—that they have one is notable—predicted that all of their employees will have avatars in five years. Some projections claim that 80 percent of active Internet users and Fortune 500 enterprises will have a *Second Life* presence in the not-too-distant future.

If present growth rates hold, the number of Internet users worldwide could triple in four years, as will their time spent online, with the largest growth occurring outside of the Western world. Certainly, more and more people benefit from virtual interaction every day, which suggests a tipping point will be crossed, as popular social venues move from physical to the digital worlds. We are at the early stages of a dramatic shift in "cyber-existence"—think of it as the difference between 2-D and 3-D, between the merely interactive and the fully immersive.

In this book, we provide an account of how virtual reality is changing human nature, societies, and cultures as we know them. Our goal is to familiarize readers with the pros and cons of the brave new world in which we live. We explore notions of consciousness, perception, neuroscience, media technology, social interaction, and culture writ large, as they pertain to virtual reality—and vice versa. We are writing for a wide range of readers—science lovers, futurists, and, most important, anyone who has a sense, somewhere in the back of their minds, that the world is changing radically as more and more of life unfolds digitally. It's thrilling, exciting, and scary all at once.

This book aims to indulge the reader's curiosity not only for whatever is just around the corner virtually, but also for the distant future. Although we sometimes use science fiction to provide colorful examples, this book is grounded in scientific theory and empirical research (much of which we conducted ourselves).

Disruptive as it may seem, the shift to an ever more virtual world—of which the Internet was only one step—may be something close to inevitable, given how humans are wired neurophysiologically. Driven by imaginations that have long sought to defy the sensory and physical constraints of physical reality, humans continuously search for new varieties and modes of existence, only this time we're doing it via the supposedly cold machinery of digital space.

CHAPTER ONE

DREAM MACHINES

ANY BOOK ABOUT VIRTUAL REALITY HAS TO START WITH A DEFINITION of what reality is in the first place. Given that philosophers have wrestled with this subject for millennia, it's not a simple task. Take the question *What is real?* Merely by asking it, we suppose there must be some things we experience that are, in fact, not real. For some, this is an obvious point. My kitchen table—that's real. The Greek god Zeus? Not so much. Real, not real . . . end of story.

Well, not so fast.

WHAT IS REALITY?

For humans, reality is, strictly speaking, constructed by minds. Many scientists, writers, and philosophers, such as Aldous Huxley, and even religious gurus like the Dalai Lama, have argued that all perceptions are actually just hallucinations, and idiosyncratic ones at that. Scientists know that what people see, hear, touch, smell, and taste are really impoverished versions of external stimuli. We know, for example, that there are more colors in the light spectrum than can be seen by humans, such as infrared, and more odors than can be smelled, such as carbon monoxide. Furthermore, the qualities of sensory stimuli that people perceive, such as the color of the sky, the smell of a rose, the feel of sandpaper, the sound of a low C on a piano, are not necessarily the same for everybody.

That people differ in their perceptions of sensory stimuli is indisputable. Nearsighted individuals see faraway objects less clearly than farsighted ones, and vice versa. Anosmic people can't smell. Others can't hear things particularly well. Some perceptual differences are genetic (e.g., people born with a gene that makes Brussels sprouts taste bad), some are maturational (e.g., infants and older people generally have poorer vision than children, adolescents, and young adults), and some result from disease or injury (e.g., a colleague of ours lost his sense of smell after falling and cracking his skull). Forget walking a mile in someone else's shoes, just move an inch using her senses and you would perceive reality differently.

Taking all of this into account, one might concede that reality is subjective but conclude that it's still a constant for each of us individually. But, even within a single mind, reality is in constant flux. Consider red-green color-blindness, the inability to differentiate those two colors. Statistically, 8 percent of men and less than 1 percent of women are afflicted. However, everyone experiences a form of color-blindness every day. Look at the corner of a room in which the intersecting walls were painted the same color. In answer to the question *What color are the two walls?* most people would name a single color, such as "sandalwood" or "buckskin" or whatever color the walls were painted. The assumption that the walls are painted the same hue drives people's mental perception, hence the answer. However, light in the room very likely reflects off the walls differently on the way back to one's retina. Consequently, the wavelength of the light—which defines the color—coming from each wall is different. Yet, most people fail to perceive the color difference and are temporarily color-blind. The perceptual system makes the walls seem the same color, simplifying the world for the viewer, even though viewers can override this perceptual process when they consciously try to notice the different colors on each wall. Perceptionists label this phenomenon "color constancy."

Nothing is particularly unsettling about the subjective manner in which people perceive reality. Sure, we see things differently from one another, and even from ourselves from time to time, but we still manage to come to a general agreement of what's collectively in front of us and to share our perceptions with each other. Those deviating from this collective perception, people who see, hear, or feel things that aren't physically there at all, are usually labeled crazy, victims of faulty wiring, etc. "Son of Sam" serial killer David Berkowitz, who heard voices, and Nobel laureate John Nash, who had recurring hallucinations of nonexistent people, are famous examples. It turns out, however, that even the reality shared by the "normal" among us is not necessarily the same.

Many think of the famous journalist Carl Bernstein as a man who can uncover information better than most Americans. After all, he and Bob Woodward were responsible for bringing the Watergate scandal to light. His ex-wife, however, disagreed. In her semiautobiographical novel, *Heartburn,* Nora Ephron wrote about her marriage, presumably with Bernstein, and humorously described the blinding effect of refrigerator light on the male cornea. This occurs when males open the refrigerator door to look for the butter and invariably ask (to no one in particular), "Where's the butter?" Sooner or later, and with exasperation, their spouses come to look at the open refrigerator and immediately point out that "It's right there!"—unblocked and in plain view. Ephron concluded, somewhat tongue-in-cheek, that despite the butter being displayed prominently in the male visual field, the male brain cannot pick it out of the array of other visual objects.

On a more scientific note, the University of Illinois perception scholar Dan Simons has studied similar behavior, albeit common to both sexes, labeling it "inattentional blindness," drawing particularly surprising results from a series of startling experiments. His book, *The Invisible Gorilla*, takes its name from these studies.

If we were to ask you to watch a video of two teams tossing a basketball among team members, and to count the number of times each team's players passed the ball, do you think you would notice if a gorilla walked on to the court among the players? Of course, you say? Surprisingly though, chances are about one in two that you would not.

In Simons's now classic experiment, people watched videos of teams passing basketballs and counted the number of passes. As the action in the video unfolds, a gorilla (okay, a person in a gorilla suit) actually walks into the middle of the players, stops, beats its chest, and walks off the scene—taking from five to ten seconds to

The invisible gorilla in action.

do so. A minute later, the video stops. Across several experiments, roughly half (46 percent) of the viewers did not report seeing the gorilla! We've tried this same experiment in our classes at Stanford and the University of California, Santa Barbara, and literally half the class shrieks in surprise when we show them the video the second time and instruct them to look for the gorilla. This inattentional blindness, or "not seeing things that are there," is also one of the keys to sleight-of-hand magic acts. (Try it yourself at www.invisiblegorilla.com.)

While spotting gorillas may no longer be a critical skill to people outside of the jungle, subjectivities in perception can have serious consequences. For example, people with racial bias actually see the world differently. Stanford's Jennifer Eberhardt, who studies racial prejudice and discrimination, has conducted a series of experiments that supports this notion. In the context of an experiment purportedly about how the human visual system works, Eberhardt told research participants to stare at a dot in the center of a computer screen, during which time she flashed photographs subliminally of black or white faces. The participants could see a brief color flash on the screen, but they weren't consciously aware that there were pictures of people popping up in front of them, and none reported ever seeing them.

Subsequently, Eberhardt presented the research participants a series of images consisting of a few dots, and then a few more, and a few more, and a few more, etc., until they could identify the object she was trying to depict. She was studying how many dots it would take before they could recognize the emerging object. Some of the pictures were of everyday household objects, such as hand tools. When household objects were the pictures in question, the prior subliminal flashes of either black or white faces had no effect on the number of dots it took before participants could recognize

 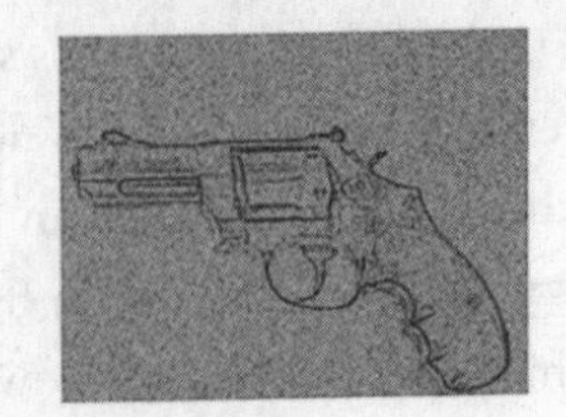 

Masked stimuli panels.

the object being depicted. However, the flashes made a big difference for the recognition of one particular object—a handgun. The participants who had been exposed subliminally to the black faces perceived the handgun reliably sooner—that is, with less available sensory information (i.e., dots)—than those who had been exposed to the white faces. Eberhardt's explanation is that the subliminal flashes of black faces had unconsciously activated violent racial stereotypes that "primed" her participants to see artifacts like a handgun, which are consistent with the stereotype. Outside of the laboratory, perhaps the priming effects that Professor Eberhardt discovered lead police to find more weapons when searching cars driven by blacks than those driven by whites, thereby reinforcing the idea of racial profiling.

Whether it's a color that we see in a particular light, a gorilla we somehow miss, or a pattern we're primed to detect, the evident variability of our perceptions undermines the common-sense notion of a hard-and-fast, fixed and static, easily defined reality. Ours is not a passive relationship, where reality *is* and we simply experience it; reality is, in fact, a product of our minds—an ever-changing program consisting of a constant stream of perceptions. And what we intend to show in the chapters ahead is how, in many ways, virtual reality is just an exercise in manipulating these perceptions.

THE PRINCIPLE OF "PSYCHOLOGICAL RELATIVITY"

Even though reality is an elusive target, most people easily divide the world between the real and the not-real. The takeaway message here is that the mind decides if perceptions are real. If the mind buys into an experience, it deems it "real," otherwise it judges it to be unreal. And, if enough people share the perception that an alternative reality is real, then who's to say it isn't? The difference between heaven, which the great majority of Americans believe is real, and leprechauns, which are fiction to most, is determined largely by consensus, as opposed to scientific proof.

To bridge the gap between what is experienced as real and what is not, let's take a cue from Einstein's work on relativity. We've parsed human perceptions into two categories, "real" and "virtual," via a concept called *psychological relativity*. Einstein introduced the modern notion of "special relativity," theorizing that the perceived speed of objects depends on the observer's own motion. In other words, what common sense tells people about movements is not always true. A speeding car appears to be moving faster if it passes you while you are stationary, than if you are driving on the highway with it. Or consider that people ordinarily believe they aren't moving when they stand still, but astrophysicists know that they *are* moving—very fast, in fact. All people move with the rotation of the earth on its axis, Earth's orbit around the sun, the solar system's orbit around the Milky Way galaxy, the galaxy's movement through the universe, and, finally, with the ever-expanding universe itself. Even on a microscopic level, subatomic particles are moving within every atom of the human body.

Analogously, the distinction between real and virtual is relative. Humans contrast what is usually considered "grounded reality"—

what they believe to be the "natural" or "physical" world—with all other "virtual realities" they experience, such as dreams, literature, cartoons, movies, and online environments such as Facebook or *Second Life*. This contrast allows us to avoid being mired in the unending debate over what constitutes reality.

As we've already seen, people often experience and believe the illusory to be real. In the drawing below, for example, line A seems longer than line B, even though both are equal objectively. It takes only a moment's reflection to realize that the distinction between "grounded" and "virtual" is often arbitrary—humans move between them.

A century ago, a classic experiment illustrated how easily our minds transport us into virtual reality. In 1896, George M. Stratton, a professor at Berkeley, described his experiences using "prism glasses." Professor Stratton wore eyeglasses with prism lenses that were designed to optically invert the physical environment, turning everything he viewed upside down. Stratton reported that he wore prism glasses from three P.M. until ten P.M. on the first day, then took them off in the dark and blindfolded himself before going to bed. Upon awakening about nine thirty the next morning in the dark room, he took off the blindfold and put the prism glasses back on, and wore them again until ten P.M. Similarly, the third day, he wore the prism glasses beginning at ten A.M. for two more hours before ending the experiment. Amazingly, on the first day, his perceptual system adjusted by re-inverting his view—even though the glasses were sending him images upside down, his brain interpreted the signal as right-side up.

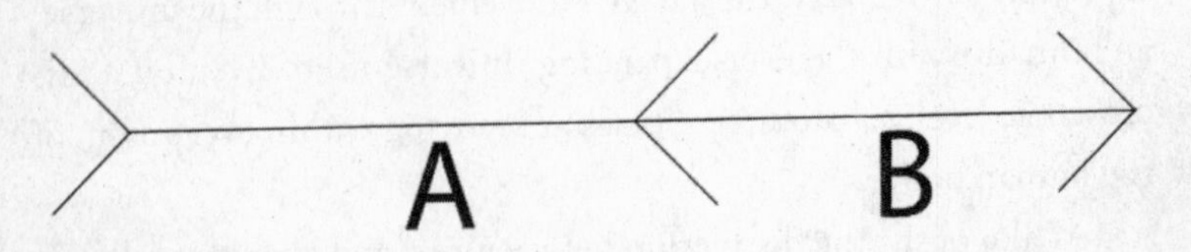

The Müller-Lyer illusion.

Many studies have since replicated Stratton's introspective report. Indeed, such experiments have become common enough that one can find prism glasses available for purchase on the Web. At first, research participants generally see the physical world as if they are standing on their heads. However, in an unexpectedly short time, their perceptual system adjusts and they see the world as they would without the lenses. When they take off the prism glasses, the grounded world, which, by definition, is right-side up, actually appears upside-down for a while. Don't worry though, eventually the participants adapted back to life without the prism glasses and saw the grounded world right-side up again.

The point here is that humans are neurophysiologically wired

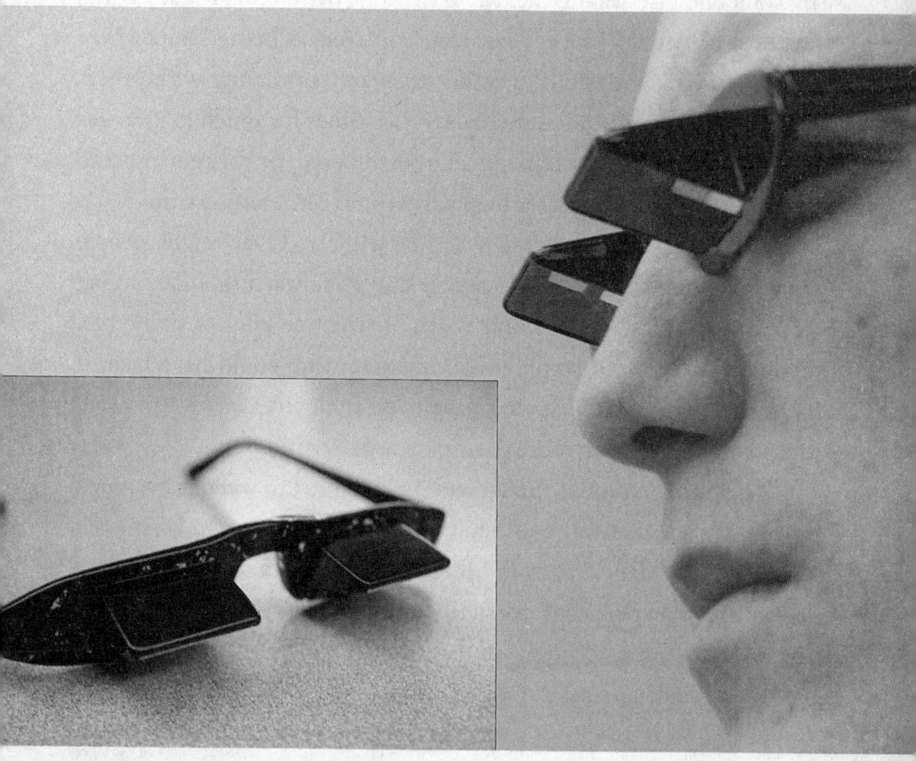

A tool to turn the world upside down.

to subjectively "right" sensory stimuli according to previously established expectations. We hope you're beginning to appreciate why people engage so readily in virtual worlds.

STUMBLING INTO VIRTUAL REALITY

"Virtual reality" typically conjures up futuristic images of digital computer grids and intricate hardware. But we believe that virtual reality begins in the mind and requires no equipment whatsoever. Have you ever spoken face-to-face with someone whose mind wandered off? Have you been startled out of your own mental reverie by someone else waving her hand in front of your face and asking, "Where are you?" Indeed, everyone experiences being "somewhere else" in their own minds, whether they are conversing with others or not—at times for a few seconds, other times for much longer—as our minds wander amid imagined, remembered, or misremembered places. Anyone who's sat through a boring meeting knows this.

Hungry people imagine what they'll eat. Overweight people imagine being skinny. Single people crave the warmth and comfort of a soul mate. Married people yearn for the freedoms of single life. People who've acted stupidly will imagine what could have been if only they hadn't. People who have lived their lives cautiously wish they would have been more daring once in a while.

The human mind is fundamentally designed to wander. So much so that it's actually more difficult to think of nothing, to keep one's mind totally "blank," than to fill it with thoughts and fantasies. Unless you're an expert yogi, you'll likely fail if you try to think of nothing, even for a few seconds.

Daydreaming—or, as cognitive psychologists label it, "mind-wandering"—is a topic of scientific research in its own right. At the

University of California, Santa Barb
collaborators have developed metho
"zone out," how often their minds wan
and, alternatively, how often people "tune
actually consciously try to "mind-wander." As
minds zone out 15 to 25 percent of the time and
wander on purpose—25 to 35 percent of the time.

To examine mind-wandering in daily life, experime
psychologist Eric Klinger and his colleagues exhaustively ex
people's thoughts throughout the day. Participants in these stu
were "beeped" randomly, and asked several questions regarding the
thoughts at the time of each beep. Combing through this vast dataset of private thoughts, Klinger's team estimated the frequency of people's mind wanderings to be about two thousand times per day!

Everyone's mind travels during sleep when they dream. But the brain mechanisms behind dreams remain a mystery to psychologists. Sleep researchers report that, on average, people experience four to six dreams per night. Like mind-wanderings during the day, dreams vary from pleasurable to nightmarish. These dreams are not just recreation—they are essential to mental health. Research has shown that the absence of dreaming is associated with physical and mental illness. For example, a major consequence of sleep apnea is diminished REM sleep; that is, dreaming.

Consider the typical undergraduate who decides to make a hundred dollars in beer money for participating in a psychology study on sleep—let's call him Greg. Greg gets paid to sleep while equipment monitors various systems, including his heart, lungs, and brain waves. He arrives at eight P.M. to be prepped, and after changing into night clothes in a private dressing room, signals his readiness to a technician, who begins attaching sensors to Greg's body. First, three heart sensors are attached—nothing to it. Next, a couple of bands are

around his upper and lower chest to monitor respiration—
so bad. The technician asks Greg to sit on a chair and begins
re arduous process of pasting many small metallic electrodes
scalp. He then tapes a small wire with a little silver disk on it
t under his left nostril, and asks Greg to stand up. He asks Greg
hold the wires from all the sensors, and helps him get into bed. It
s already nine P.M.

Greg receives instructions to relax; he can go to sleep anytime. The lights are turned off, save for a dim ceiling lamp directed at Greg's head. He meekly asks that the annoying overhead light be turned off. But the lab tech says that he can't do that, explaining that the light is needed so that the camera in the corner of the room can record his body movements during the night.

"Are you ready?" asks a voice over the intercom. Greg replies in the affirmative and, after a couple of hours of tossing and turning, he finally falls asleep. Just as he begins his first dream, something about white sand and sunshine, he is shaken awake. Groggily, Greg asks what's going on. The technician explains that the experimental protocol requires him to wake the participant up at various times during the night as part of the study. Greg mumbles, "Can I go back to sleep?" The tech nods. A while later, just as Greg starts to dream again (in a field, this time, but still lots of sunshine), there's the shake—waking him up again. Damn! The tech lets Greg go back to sleep, but again and again, Greg is awakened every time he starts to dream. In the morning, Greg is quite ready to leave, and quite tired despite actually having slept for a total of eight hours. He collects his payment and goes off feeling not quite right.

Greg just participated in a dream-deprivation study. The results of such studies suggest that dreams play a major role in sleep efficiency. Without dreams, people do not feel rested after even a full night's sleep. It seems it's not so much the hours of sleep that are important, health-wise. Rather, truly healthy sleep actually requires mind-wandering, even if one rarely remembers those dreams.

So, we're designed with wandering minds. What does any of this have to do with virtual reality? Plenty.

After all, it's clear we can be in two places—one physically and one mentally—simultaneously, whether awake or asleep. And, moreover, our mental location can actually override any sense we might have of our physical location. Mind-wandering during both sleep and wake appears to be healthy. Sometimes people tune out by consciously traveling from grounded to virtual reality. More often, they are unconscious of the transition, as when they stumble (zone out) into dreams and daydreams.

Humans have toyed with and discovered numerous ways to facilitate this mind-travel for tens of thousands of years. One class of such tools is pharmacological.

REALITY UNDER THE INFLUENCE

The Internet is the LSD of the 1990s.

—Timothy Leary

Virtual reality experiences do not require any media technology. Indeed, for thousands of years, humans have discovered and used psychoactive substances found in nature as well as manufactured pharmaceuticals. So-called hallucinogens facilitate mind-wanderings—"trips"—to virtual realities by directly affecting the brain and the nervous system. Hallucinogens like peyote cause changes in perceptual, cognitive, and emotional experiences. These drugs can be addictive not because of a physiological dependency but because of the favored quality of the virtual experience they produce, relative to one's life in grounded reality. Psychedelic intoxication is often regarded as a mind-expanding experience that allows conscious exploration of alternative (i.e., virtual) realities. But as a good friend of ours used to say, "I've been on dozens of acid trips, but not a good one yet!" Indeed, the result of this transportation can be brutal if the destination is one of paranoia and fear.

Drugs are also sometimes used to bring people *back* to what society defines as grounded reality. For some individuals, hallucinations are a normal, waking occurrence. Recall our brief mention of "Son of Sam" killer David Berkowitz and mathematician John Nash, who frequently hallucinated the presence of others' voices, bodies, or both. For such individuals, anti-hallucinogenics such as Stelazine or Thorazine are used to reduce hallucinations so that they can share the same grounded reality as most other people. Paradoxically, for such individuals, society's grounded reality is their virtual reality. It's a fascinating thought to ponder.

In sum, people have used various substances for thousands of years to facilitate what we call virtual experiences. Humans, it seems, simply enjoy taking a vacation from grounded reality.

THE ULTIMATE DESTINATION

Historically, virtual reality is perhaps most commonly found in *religion*. Most people profess to be religious, and almost all religions espouse the existence of an eternal, divine reality that is superior to the temporal world of the flesh. Many promote stories about an afterlife. These stories teach the faithful that what they assume is grounded reality—the physical world around them—is really only a temporary virtual world that will fall away after death.

For example, in Christianity and Islam, grounded reality is an eternal divine paradise that awaits only those who keep the faith of their religion, and in Judaism many believe the dead will be resurrected into the "World to Come." These religions differ in the details of how the afterlife proceeds, but are all in agreement that the corporeal life we live is but a moment on the eternal calendar—a virtual life, perhaps created to test or prepare us. The world that houses our physical body is a mere temporal stopping point on the journey to the real, eternal, grounded world.

FAR FROM BEING AN ODD HOBBY OF GEEKS, IT WOULD APPEAR THAT from our species' earliest days virtual reality has been a large part of the human experience. Stray thoughts, dreams, chemical influences, or religious beliefs all point to an inherent human need and desire to transcend reality "as we know it." There is nothing inhuman, in other words, about the virtual. Of course, for our purposes, the most important virtual tools humans have developed, and have had for some time, are media-based. These are the tools that have allowed an explosion in individual and, crucially, shared virtual experiences over the years. They are our focus in the next chapter, and are the driving force of the rest of this book and the future of virtual reality.

CHAPTER TWO

A MUSEUM OF VIRTUAL MEDIA

PEOPLE TRAVEL VIRTUALLY VIA MEDIA ALL THE TIME. THIS IS NOT to say there are no other important functions of media—for example, communication and learning—but throughout human history, people have put media to work to enhance mind-wandering. For one thing, if you're on a quest for transportation into a virtual world, media tends to be safer than narcotics.

If a hands-on Museum of the History of Virtual Reality Technology existed, visitors could examine milestones in communication technology—each a new way of transporting the mind from grounded to virtual reality. Imagine yourself in such a museum, with the exhibit rooms roughly in chronological order from the distant past to the present. With each exhibit, think of the tools and technologies people have created to facilitate virtual travel.

EXHIBIT 1: STORYTELLING, THEN AND NOW

In the first of this two-room exhibit, you see an obvious spot on the floor marked VISITOR, for you to sit. You do so, finding yourself in a group of what appear to be Neanderthal men and women. Even though they're animatronic robots, you become entranced as one of them stands up and relates a story about where the animals surrounding the nearby village originated, how an all-powerful spirit put them there. You're fascinated and get caught up with the robotic cave-people, who seem to be listening to the story.

In the second room, distinctively modern in décor, you observe several museum visitors sitting on couches arranged around a working fireplace and listening to a radio. In his weekly radio show, *A Prairie Home Companion*, Garrison Keillor is melodically mesmerizing everyone with his descriptions of the people and happenings in the fictional Minnesota town of Lake Wobegon, "where all the women are strong, the men are good-looking, and the children are above average." You listen for a while, feeling that you're getting to know the good citizens of Lake Wobegon. Indeed, as you close your eyes and let the story envelop you, it seems you can smell the stale beer at the small town's Sidetrack Tap.

EXHIBIT 2: GRAPHICS

Here, you learn that even for ancestral humans, words were not enough—so graphics were invented. The room is a large cave. At the entry, a sign informs you that the cave is a replica of one in Lascaux, France, where cave paintings date back at least thirty thousand years. The cave walls are covered with several primitive drawings, some of large animals in motion, some of hunters, and some of curvaceous women. You sit on the cave floor as three archaeologists debate the meanings and functions of this primitive art. One argues that the paintings were made by ancient shamans, and she claims that shaman painters went into caves where they would self-induce trance states, perhaps with the aid of drugs, and paint images of their mental visions to share with those who would view them later. The second archaeologist argues that the ancient cave paintings revealed the fantasies of adolescents who made up most of the male population way back then. He points out as proof that their "graffiti" expressed powerful animals, hunting scenes, and highly sexualized depictions of women, themes also common to many motion pictures today. The third argues more mundanely that cave paintings, most of which involve animals, were created by early humans to magically increase the prospects of hunting success.

Suddenly, the cave darkens and a fur-wrapped cave-dweller enters with a large burning torch, which he plants in the soft floor of the cave. The fire casts his shadow on the cave ceiling. He entertains everyone with a series of hand and finger motions creating a panoply of moving shadow animals. You realize that this is likely the first animation technology.

EXHIBIT 3: SCULPTURE

The next exhibit reminds you of the sculpture garden at your local art gallery. The date and place labels show that sculpture emerged in many places throughout the ancient world including Africa, Europe, Southern Asia, East Asia, and the Americas at roughly the same time as cave-painting. For example, the "Lion Man," the oldest known statue that depicts an animal, was found in a cave near Ulm, Germany. Scientists believe it was created over 32,000 years ago! You also notice that ancient sculpture seems to have been more closely tied to religion and the gods than the ancient cave-paintings. Indeed, a descriptive plaque relates that many sculptures in fact were believed to be the physical manifestations of the gods—in some cases, the gods themselves. You recall a recent news story about some Christians who believe that saints inhabit statues that weep and bleed, despite the doubts of many, including religious leaders.

EXHIBIT 4: THEATER

You arrive at a large exhibit, a re-creation of an amphitheater in ancient Greece, around 650 BCE. A play is unfolding on the stage far below. Given the complexity of the performance, compared to the earlier exhibits, you realize that stories could be recorded in scripts and passed on to later generations. People's imaginations could be stimulated via written stories (scripts), painting (scenery) and sculpture (props), and actors simultaneously, allowing audiences to more easily zone out of everyday reality and enter virtual reality. Ancient playwrights throughout the world produced stories that unfolded dynamically (i.e., over time), complete with 3-D graphics (i.e., props), with great success. You recall once really getting into character, perhaps virtually becoming Puck in your high school's production of *A Midsummer Night's Dream.*

EXHIBIT 5: MANUSCRIPTS

The next exhibit is arranged chronologically. In the first display, you learn about the invention of ink, said to have occurred about 1500

BCE, and paper in the second century CE—both in China—as well as the quill pen in about 700 CE in Europe. You also learn that these inventions facilitated the transmission of knowledge via scrolls and manuscripts. You see a diorama of lifelike European religious monks busy at work at an important aspect of their calling, preserving religious tracts for the future by hand-copying and decorating them. The archives produced with this medium allowed for widespread and sustained virtual travel.

EXHIBIT 6: MOVABLE TYPE

You know from the prominent exhibit sign that this one is a "must-see." You learn that although printing using hand-carved blocks of wood was practiced as early as 200 BCE in China, it was generally used only to print illustrations, as there were far too many unique Chinese symbols. You accept the docent's argument that the most important communication media technology invented since writing was movable type, individual alphabetic characters that made the modern printing press possible. The mid-fifteenth-century invention was credited to the German goldsmith Johannes Gutenberg. In the tradition of Western European monks, the first major work that Gutenberg printed was, appropriately enough, the Bible, the Western world's best-selling book. You marvel at a copy through a sealed glass case.

Also exhibited here is a complete set of J. K. Rowling's Harry Potter books, with over 500 million books (translated into more than sixty-five different languages) sold as of 2010. In aggregate, this series is likely the next-best seller (to the Bible and the Koran) of all time! You remember how most children and many adults were enraptured by the saga. Maybe you recall watching the first Harry Potter movie with your daughter, who exclaimed of the fictional Hogwarts Academy, "It looks just like it's supposed to look, it's really real!"

EXHIBIT 7: PHOTOGRAPHY

In the anteroom, the docent sheds light on the obvious, that photography was an important milestone in the history of virtual media technologies. You see many of photography's technological forerunners up close, including early pin-hole cameras and the camera obscura. You learn about how chemicals interact with light to form photographs. In the next room, you see a copy of French inventor Joseph Niépce's first photograph, developed and printed in 1825, and try to guess what exactly is in the picture. Is it a man at a drafting table or a scene of a French village? Indeed, even when media try to capture grounded reality, there is room for interpretation.

A copy of French inventor Joseph Niépce's 1825 photograph. What do you see?

You read that the people of Niépce's day were amazed by his invention and that, according to photography historian Mary Warner Marien, people's reactions to early photographs "ranged from the exuberant to the cautious." There is a quote on a wall from writer Edgar Allan Poe, who, apparently smitten by photography, wrote, "[A]ll language must fall short of conveying any just idea of the truth . . . but the closest scrutiny of the photogenic drawing only a more absolute truth, a more perfect identity of aspect with the thing represented." What an amazing admission for a writer quite capable of scaring people virtually out of their wits with his short stories!

You realize how startling photographs would have been to the cave people you joined in the earlier exhibits, who might have cowered or perhaps attacked the photograph. You see the first color photograph, "Tartan Ribbon," taken by renowned physicist James Clerk Maxwell in 1861. It reminds you that it's time to scan all of those family snapshots and save them electronically for posterity.

EXHIBIT 8: CINEMATOGRAPHY

In the next room, you realize that the history of media technologies leading up to photography was paralleled by an equally long history of attempts to represent the movements of animals, people, and objects that stretch as far back as cave paintings, which often were "stills" of animals in motion. Even today, you recall, many art critics speak of "movement" in still paintings.

The first stop of this exhibit is a pad of paper with a drawing on the top page. Following the instructions, you rifle the stack from the lower left corner. When you do, stick figures that were previously standing still dance as the pages fly by. You soon learn that movies are conceptually similar to flipping through such pads of paper. Instead of drawings, sequential photographic negatives are used. And, rather than manually producing the "motion picture" effect, a reel of film containing still photographs is mechanically routed past a light source and projected on a screen.

You're told that modern movies began appearing in the late nineteenth century, and that the first screening of a movie for an unfamiliar audience is often credited to Auguste and Louis Lumière, French brothers who in 1895 premiered a short film titled *The Arrival of a Train at La Ciotat Station*. The reel plays, and you see that the camera shooting the film was placed very near the railroad tracks and captured a train's motion as it approached, leaving the audience with the view of an oncoming train rushing toward them. Apparently, the sensation brought on by the film proved unbearably scary for many viewers in the audience, and some were said to have fled the building in which the movie was screened.

Why such hysteria? The docent explains that the viewers weren't

A still from The Arrival of a Train at Ciotat Station.

accustomed to viewing movies, but they were used to seeing trains. They knew how big trains were and how messy it would be to be hit by one. They knew consciously that they couldn't be physically injured. But part of their minds reacted as if the trains were real, inducing what we call involuntary virtual travel. Viewers could not consciously control their automatic, unconsciously driven fear responses.

EXHIBIT 9: ELECTRICITY

It doesn't seem like an example of communications media, but you stop by this room to hear a scholar talk about the history of electricity and its role in media. He speaks of how the discovery, production, and transmission of electricity revolutionized motion pictures, recorded music, and communication. Unlike prior media technologies, electrified media made it possible to communicate in near real time over long distances and to multitudes of people simultaneously.

You learn that electricity and the telegraph wiped out the Pony Express only eighteen months after it began. Samuel Morse's telegraph (the first transcontinental lines were completed in America in 1861) provided for coded two-way communication in real time that could be transmitted from East to West and back again. Instead of taking days to cross the country, it took seconds. A half-century later, telephone lines incorporating amplifiers linked Alexander Graham Bell's voice telephone coast-to-coast. Today, despite huge advances in all sorts of media technology, the "telegraph model" still is quite a successful one, as can be witnessed by the rampant use of text-messaging among young people.

EXHIBIT 10: BROADCAST MEDIA

You move on to the next exhibit to discover that electricity begat electronics in the late nineteenth and early twentieth century, in the form of vacuum tubes. The invention of the first practical vacuum tube is credited to Ambrose Fleming in 1904, and by 1920, vacuum tubes were being manufactured by the RCA Corporation. Indeed, vacuum tubes provided the amplification technology for long-distance phone lines.

You learn that vacuum tubes also made short-wave and commercial radio possible. People could now broadcast sounds simultaneously to very large audiences, and also had an alternative to the telephone for two-way audio communication. Later, the invention of television made possible the multimedia combination of audio and video transmission. Radio and television are quite familiar today. Just before exiting, you stop to don a pair of glasses and watch the last six minutes of the most recent NFL Super Bowl on 3-D television. It's like being on the field.

EXHIBIT 11: THE COMPUTER AND INTERNET

Vacuum tubes were used to construct the first digital computers in the early 1940s. In 1946, the first general-purpose electronic digital computer, the ENIAC ("electronic numerical integrator and calculator"), was constructed at the University of Pennsylvania. The digital age had commenced. The invention of the transistor revolutionized electronics, replacing inefficient vacuum-tube-based computers with solid state-based digital computers. Ultimately, in the late 1970s, microcomputers (the forerunners of personal computers) appeared on the scene.

The electronic numerical integrator and calculator (ENIAC).

Software advances accompanied the miniaturization of computers. Word processing became commercially available in the late 1970s and early 1980s, a feature that made computer users out of millions and millions of people around the world. Almost immediately, in-house local networks made the first e-mail capabilities practical, and such systems became the envy of corporations without them. Soon, however, demand drove the establishment of a network of networks, and the beginnings of the Internet began to take form. It grew fairly rapidly and led directly to then senator Al Gore's "information superhighway" bill that made the Internet possible, evidence that the politician played a role in its development.

By the early 1990s, the Internet as we know it today was developed. It revolutionized human communication. When President Clinton was inaugurated, there were only about fifty Web sites. Today, that number is so high it can only be estimated. Advances in computer hardware and software graphics accompanied the Internet's growth.

The Internet is now the main forum for socializing, commerce, and entertainment. On the Internet, long-lost friends and relatives are found by searching for names and identities. Relationships are fostered by "smart" online dating sites that match personalities, and then dashed by spouses reading about each other's transgressions online. The Internet allows for grass-roots democracy with the flow of information that was previously hard to come by, while also spotlighting countries that censor Internet searches. Fortunes were made in "dot-com" corporations and also lost in offshore gambling Web sites. The Internet may define the generation growing up in the current era.

EXIT

Your day at the museum ends. You reflect on the sweeping arc you've just traveled across human civilization and realize that, all along, we've combined arts and science to produce mind-wandering tools. Virtual reality is nothing new. It's just a new label.

ADVANCING VIRTUAL REALITY

We believe that we are at or slightly past the threshold of creating technologies that, for good or bad, will transform the experience of being human. Life will be seamlessly altered, if not enhanced, as digital technology becomes part and parcel of our daily lives, allowing humans to break through the constraints of past technologies.

Consider the telephone. When asked with whom one is conversing "on" the telephone, one doesn't say, "That was Joe's digitized voice." One says, "It's Joe!" Indeed, most people are totally unconscious of the simple fact that the voice they hear during a cell-phone conversation is not the other person's voice but a digital construction that only approximates that voice and is measured and produced nearly in real time. The fidelity of that voice is simply good enough that our conscious powers remain quiet, never injecting doubt into our minds about the reality of the person with whom we're interacting.

Technology will soon provide levels of visual fidelity (and, later, touch and smell) comparable to the audio fidelity of the telephone. In other words, virtual worlds encompassing all of the senses (roughly speaking, think *The Matrix*) will at some point "feel" as "real" as a telephone conversation does today (imagine *Avatar*). Might those worlds be disorienting at first, like the prism glasses and their upside-down world? Perhaps. But can and will we adapt in short order? Absolutely. Now that the technology has begun to catch up to the sci-fi musings of the past, we can begin to see how humankind will soon ground more and more of its reality in virtual worlds—to astonishing effects for existence, individual and collective. In a visit to our museum three hundred years down the road, the virtual-reality technology exhibit (not yet built during our visit a few pages ago) may represent a quantum leap of sorts, a turning point in our history distinctly marking what came before and after it. So, consider the rest of this book to be that very exhibit. Welcome to the next great advance in the history of human communication.

CHAPTER THREE

MIRROR, MIRROR ON THE WALL

ONCE A YEAR, MATT GROENING'S COMIC STRIP *LIFE IN HELL* FEATURES "Forbidden Words," those phrases uttered so many times that year, one can hardly bear to hear them again. Examples from previous years have been "guesstimate," "I'm all like . . . ," and "[insert any color] is the new black." These days, "virtual" might make it to Groening's list—everyone seems to be dropping it. Slap "virtual" into a business plan, and entrepreneurs attract copious venture capital. For the remainder of this book, unless stated otherwise, when we use the term *virtual reality*, we generally refer to the digital technology–based immersive virtual reality that we describe in depth in this chapter.

In the history of virtual media, some inventions have cracked open the departure gates just a bit. For example, today's high-definition broadcast television is more entrancing than its predecessors. Every once in a while, however, a new technology blasts the gates off their hinges. Movable type and the printing press come to mind. We believe virtual reality–based digital technology fits the latter category.

THE VIRTUAL PIT

In virtual reality, gravity is optional. People can walk on air. However, scientists choose to keep gravity "turned on" to prove just how compelling virtual reality is.

At just about every virtual-reality laboratory we've ever visited, scientists proudly and sometimes deviously roll out their version of a "pit" demonstration, which "wows" visitors curious to experience virtual reality. This demonstration is particularly effective for people initially skeptical about how "realistic" the experience can be. The basic scenario is simple. The skeptic dons a clunky helmet, a *head-mounted display*, which limits his visual field to 3-D images displayed on miniature screens. These digitally created images induce an acrophobe's worst fear.

During a pit demo, someone might find himself in a nicely furnished virtual room when, suddenly, a section of the floor lowers like a fast elevator, revealing an opening with a deep drop to the floor of a room below. Everything else "upstairs" remains static. Looking over the edge of the pit, nearly all visitors experience anxiety. Their fear is exacerbated when a "wooden" plank spanning the pit appears and the visitor is asked to walk across it, balance-beam style. Typically, their toes curl, palms sweat, and heads shake "no." If the visitors are brave enough (and we've found that about one in three adults are not) to "walk the plank," they struggle to maintain balance. If they "fall" into the virtual pit, they sometimes gasp or even shriek in fear, often bending their knees to cushion the even-

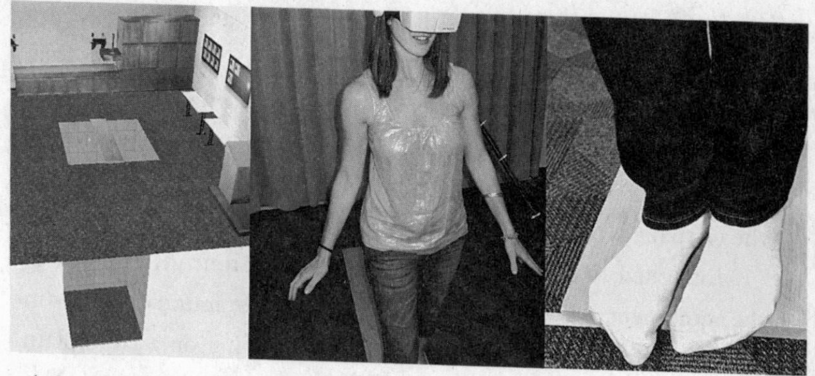

A visitor to the virtual pit watches her step.

tual landing. Of course, there is no pit in the physical world in which the demo takes place, so, unfortunately, visitors sometimes wind up crumbling to the floor (which is why we hire spotters, whose job it is to catch the visitor for such demos). On a few occasions, visitors have been so frightened that they've even sprinted away from the edge. (Unfortunately, doing so in a physical room of limited size can lead to some bruises.)

Indeed, it was an unplanned walk over a virtual plank that first caused us to consider studying the psychological aspects of virtual reality. In the late 1990s, we were walking the hallways of the psychology department at the University of California, Santa Barbara. Passing by the open door of Professor Jack Loomis's lab, we observed a young graduate student outfitted in an obviously high-tech sort of helmet and an elaborate backpack seemingly straight out of Dr. Emmett Brown's garage in *Back to the Future.* Just by watching her move, without knowing what she was seeing, we understood clearly that her mind was somewhere other than where her body was. At first, it looked like she was playing blind man's bluff. However, soon her head and body movements hinted that she was looking for something. She walked deliberately but carefully, making right and left turns now and then. Our intuitions proved correct. Indeed, it turned out that, like a mouse in search of cheese, she was negotiating a virtual maze and trying to find an object. Her head and body movements were obviously intentional and driven by the digitally created contents of the virtual environment in which she was immersed.

Intrigued, we asked Jack to give us a virtual tour. After donning the equipment, we waited a few seconds. We then found ourselves in a rather sparse 3-D virtual environment, standing on a large plain with a bluish-gray sky above. Jack directed us to walk around. Not much to see, not much to think about, but we could move freely and naturally within it. If it hadn't been a new experience, our minds might have wandered.

However, Jack instructed us to look down. A kind of sinkhole (common to battle scenes in *Star Wars*, rural locales in South Florida, and virtual-reality laboratories) opened up close by on the plain. It appeared to be about three meters square and was deep enough so that we could not see the bottom. Jack suggested that we walk right up to the edge of the abyss and look down. Being careful not to fall in, we estimated the pit to be about ten meters deep. Suddenly, a precarious wooden plank appeared, bridging the pit. Jack suggested we walk across it. No way! With a little more encouragement, we eventually walked the plank, though gingerly.

Once accustomed to our new environs, our minds did wander for a brief moment. Turning on our researcher-modes, we remembered that we were actually in Jack's lab. We knew there was no physical pit corresponding to the virtual one, and that there was little chance of injury from falling, but just like the audience of the Lumière brothers' train film, we couldn't consciously control our fear response. We began to realize that the mind can't easily be in two places at once.

Afterward, back in the physical world, we realized what a big boon virtual reality could be for the study of behavior. The time had come when experimental scenarios in virtual worlds could appear very much like their physical-world counterparts.

For more than a decade, we have run thousands of people (ranging in age from six to eighty-four) in pit demonstrations. One, in particular, stands out. We met him in 2002, at a demonstration for about a hundred judges, lawyers, and government-policy officials at the Federal Judicial Center in Washington, DC. We were leading a session on how virtual reality could be used to improve the legal system via, for example, virtual police lineups and crime-scene re-creations (more on this later). Part of making our case was demonstrating that virtual reality could be psychologically compelling enough for legal purposes. For starters, we turned to the dreaded virtual pit.

One official became quite deeply immersed very quickly. He accidentally "fell" from the virtual plank and, realizing that he was about to "plummet" toward the bottom, physically lunged horizontally, desperately trying to grab the opposite edge of the virtual chasm to "hang on" for dear life. To our dismay, his mind abruptly returned to the physical world when he did a face-plant on the marble demonstration-room floor. We were terrified that this incident would delay any use of virtual reality in courtrooms for a long time. Luckily, he was fine, and our fears were allayed. The only damage was a slight bruise to his ego.

Not even we are immune to such powers of virtual reality. In 2006, we took a trip to the University of North Carolina at Chapel Hill to visit one of the great pioneers of academic research in virtual reality, Fred Brooks. Professor Brooks has watched virtual reality grow for the past forty years. Born in 1931, he received his Ph.D. in applied mathematics from Harvard in the 1950s. At eighty, he still runs one of the most technically advanced virtual-reality labs in the world, a facility that occupies nearly an entire floor of the computer science department at UNC and is staffed by a small army of Ph.D. students and postdoctoral fellows. Indeed, he has been building virtual pits since the early 1990s.

Professor Brooks was kind enough to demonstrate his pit to us. Suffice it to say that Dr. Brooks really knows how to run a virtual-pit demo. When we initially donned the equipment, there was zero fear in our hearts and minds. After all, we had built many virtual worlds ourselves. But the student giving us the demo did something strange. She asked us each to hold a napkin in the palm of our hand. A glance at Professor Brooks revealed a knowing look.

When the simulation finally appeared, our toes were atop a virtual plank, which was par for the course. But this time we were absolutely terrified. Our hearts beat hard and fast, and it took every ounce of courage to make ourselves traverse the plank safely—after all, we couldn't have one of our scientific heroes think that we were afraid of a standard virtual-pit demo. Afterward, Brooks asked us to examine our napkins. We had sweated through them, and he had made his point. There was very little about Fred Brooks's pit that was "only virtual"—even to experienced hands like us. So, what was it that made his pit so much more compelling than those in our labs, which had compelled nearly everyone in their own right? It had nothing to do with color, or resolution, or anything about the graphics. One key to making virtual-reality immersive is a technology known as *tracking*, and the tracking in Dr. Brooks's lab was spectacular.

In the following section, we unpack the technical aspects of how virtual reality functions. In particular, we focus on three concepts: tracking, rendering, and display.

TRACKING AND RENDERING

Atari's *Pong*, one of the first commercial video games, became a hit in arcades, bars, and living rooms in the late 1970s. *Pong*, a rudimentary form of table tennis, was a simple video game by today's standards. The game consisted of a cathode ray television monitor and a dial controller. Players rotate the dial clockwise or counterclockwise to move a "paddle" along a single dimension, using it to both defend against, and carom a "ball" back across the screen in an attempt to get it past the other player. Baby boomers became as addicted to this game as their kids are to modern ones—indeed, according to Allan Alcorn, one of the creators of *Pong*, the first service call for a *Pong* arcade machine was necessitated by an overstuffed coin slot.

Readers may be more likely to recall *Pac-Man*, in which a player controls a "biter" (a sort of circle with a chomping mouth) using a joystick to navigate a maze in two dimensions (vertically and horizontally). Despite its still-rudimentary tracking, *Pac-Man* became a cultural phenomenon, with "*Pac-Man* fever" infecting popular songs and television programming.

Today, Nintendo's Wii for Xbox, Microsoft's Kinect, and Sony's PlayStation Move are all the rage. Wii players use a wand with inertial sensors to control their avatars and equipment such as golf clubs, bowling balls, baseball bats, and tennis rackets—indeed, Wii's tennis game is the modern analog of *Pong*. The key difference is that

Examples of video games, from less immersive to more immersive.

players control tennis rackets by swinging their arms in three dimensions (i.e., vertically, horizontally, and in/out), as though they are actually playing tennis. They see their virtual arm moving more or less in synchrony with their physical arm on a monitor or projection screen—the bigger the better. Kinect players don't even need to use the wand! A special camera figures out how their arm is moving and moves the avatar arm accordingly.

These examples illustrate the concept of *tracking*, that is, measuring a user's physical movements. *Pong* tracks movement in one dimension, *Pac-Man* in two. The Wii, Move, and Kinect provide even more. As the tracking becomes more intricate, these games become more immersive.

IN ADDITION TO UTILIZING TRACKING TECHNOLOGY, GAMES MUST also utilize *rendering* technology to produce the images and sounds used in games. Rendering works in tandem with tracking to continuously update player views as a function of the tracked movements. In a repetitive cycle, the user moves, the tracker detects that movement, and the rendering engine produces a digital representation of the world to reflect that movement. For example, if the joystick in *Pac-Man* tracks a movement to the left, the rendering engine draws the biter moving to the left. When a Wii tennis player swings her hand, the tracking wand detects the movement and the rendering engine draws a tennis swing. The Kinect tracks wandlessly.

Sophisticated tracking and rendering technologies have been around for a long time. Although people never think about it, in principle, telephones operate as virtual audio tracking and rendering systems. These days, when a person speaks, the telephone tracks a speaker's sounds by digitizing them—converting the information to binary strings of ones and zeros. Telephones use that digital information to reproduce—or "render"—synthetic sound to the listener.

Global positioning systems (GPS) provide another common example of tracking and rendering. Far above the earth, satellites continuously track a car's position via a signal emitted by a GPS transceiver and relay its position back to that transceiver. The GPS computer calculates the latitude, longitude, and even altitude of the car's location at any given moment. This process cycles many times per second, and the location information enables the GPS computer to continuously update a representation of the car on a stored map via a digital display. It is a repetitive cycle—track the position, update the car's location, and display the updated map with the car on it to the driver.

Virtual-reality technology works similarly, but human movements are a bit more complicated. The bad news for tracking and rendering in virtual reality is that not all human body parts move or point in the same direction simultaneously, as most do in the body of a car. The good news is that programmers can safely assume that human body parts, like car parts, are attached to each other.

In virtual reality, tracking is accomplished not by a satellite but by video cameras, magnetic sensors, and other instruments that capture signals from devices worn by users. The latter devices include reflective patches, light-emitting diodes, magnets, and accelerometers, which transmit a continuous stream of data that pinpoint users' body movements. The tracking data also pinpoint where users are looking (called a "point of view") very precisely at any given moment. The tracking equipment scans users dozens of times per second to determine, for example, if a participant has changed his point of view, if a toe has moved forward, if he or she is leaning sideways, and so on.

Consider the relatively simple tracking setup depicted above. The user wears a helmet topped with an infrared light-emitting

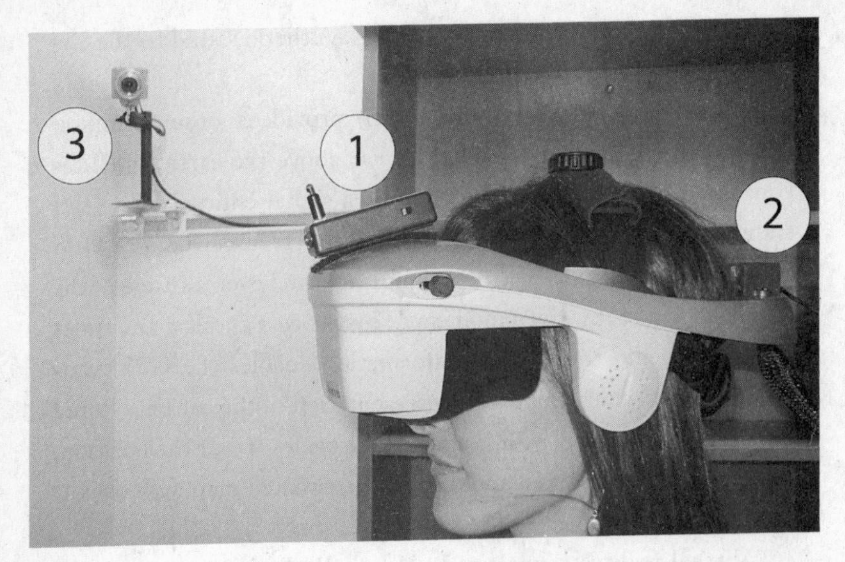

Immersive virtual-reality gear.

diode (marked 1 in the figure). A computer "watching" images captured by video cameras (3), constantly finds this tiny light. The computer locates exactly *where* (within approximately thirty millimeters) the light is in 3-D space and *when* it is there (within one-hundredth of a second). The light fixed to the top of the user's head enables the system to track where a user is at any given moment and in what direction he or she is moving. These types of movements are called *translations*.

However, translation information does not indicate in what direction the user is facing as he or she moves. It is also important to track body *orientations*, such as body turns and head turns, which are independent of changes in X, Y, Z position. Without orientation data, the system would track that a user is moving, say north to south in a virtual alley. However, it would not know if the user was walking forward or backward or whether her head was pointed straight ahead or to left or right, up or down, etc. The black cube (2) is an accelerometer that tracks orientation of whatever it is placed on. Having both gross body movement (i.e., translation) and direction (i.e., orientation) are necessary for rendering immersive 3-D virtual-reality experiences.

Tracking information is important, because the system needs to know where someone is looking—his "viewpoint" or "point of view." Much like the way a GPS computer in a car stores a map, the computer powering the virtual system stores a 3-D digital model of the virtual world—for example, spaces, such as a room or prairie, and objects in it, such as furniture and buffalo. Using the tracking information and data from the stored digital model, the computer updates the user's viewpoint by rendering the 3-D world from that viewpoint. In the pit demo, if the user tilts his head and looks down, he will see the pit. If the user turns his body and head around 180 degrees, he will see the rear wall of the virtual room. Every moment the user moves his head and eyes physically, the tracking equipment detects the movement, causing the computer to render the world appropriate to this new point of view. In a state-of-the-art system, this process repeats itself approximately one hundred times a second. To the user, everything appears smooth and as it should be.

When *Pac-Man* players want their "biter" to move to the right, they use actual human motion, moving their hands to the right with the joystick. However, even though the game is engaging, players have only an external, "bird's-eye" view, as an outsider looking down at the maze. There is little sense of being in the world itself. However, given the viewpoints that modern virtual-reality technology enables, users can move and gaze at things, much like in the physical world. For example, users approach a virtual cliff as they would a nonvirtual one. Such tracking allows for *naturalistic* movements rather than ones that are not intuitive.

What makes Professor Brooks's virtual pit so much more compelling than others is the sensitivity of his tracking devices. They capture user movements more often (almost twice as fast) and more accurately (the correspondence between physical movements and virtual ones were at the level of millimeters) than other systems.

Tracking and rendering are often difficult concepts to grasp for people new to virtual reality. We asked Jaron Lanier, who coined and popularized the term *virtual reality*, about the most intuitive way to explain why tracking is so important. He described the human perceptual system metaphorically as a "spy submarine." The brain likes to know what is going on in the outside world, so it uses information from its periscope (i.e., the eyes, ears, and other sensory organs) to move about and gather information. In other words, humans are constantly moving and updating their viewpoint to understand what is going on in the world around us. Consequently, to make a virtual-reality experience compelling, the human perceptual system needs constant updates about how the virtual world looks and sounds. In a perceptual sense, the best way to accomplish this is to allow people to move as freely and naturally as possible in virtual reality.

Just how critical is this movement? Perceptual psychologists have run experiments in which they actually freeze the muscles around the eyes, so that they can't be moved, and have discovered that if the eyes can't move, and if the body and head are fixed in place and the scene is static or frozen, then the brain cannot see anything, even if light still hits the retina in the normal manner!

Contrast the naturalistic movements in virtual reality to the artificial movements required by controllers such as keyboards, joysticks, etc., people use—for example, in handheld or desktop computer games. Such games can include 3-D digital models of spaces and objects; for example, the walls and floor of rooms and outdoor scenes, such as virtual cliffs, but if the player is merely sitting down with her head fairly stationary, the experience is far less compelling than if she can freely move her head and hands and eyes to interact with the virtual world.

ONE BENEFIT OF TRACKING SYSTEMS IS THAT THE DATA PRODUCED can be stored, analyzed, and used to reveal all sorts of important behavioral information. For example, Jesse Fox, a professor at Ohio State University, strives to understand how various media, including virtual reality, influences attitudes toward women. For example, in the extremely popular video game *Grand Theft Auto*, players can employ a prostitute and then kill her to get their money back. Given how potentially disturbing these simulations are in terms of forming gender stereotypes for the children who play them, Jesse wanted to quantify how "sexist" men are in virtual reality.

To do so, she put about 120 male Stanford students in a virtual-reality simulation with computer-generated women in various stages of, well, undress. While the men walked around the virtual room, the system constantly tracked their gaze—that is, their point of view. Some of the women were dressed conservatively; others were more scantily clad. The stored tracking data allowed Fox to construct the graphs depicted on the next page, which reflect the gaze of two typical male participants—one looking at a "sexualized," scantily clad woman in the room (left panel) and one looking at a conservatively dressed woman. Using every tracking sample taken—one every sixtieth of a second—Fox illustrated where the men were looking. The density of dots makes it a simple matter to determine where the male eyes focused. Which of these two tracking patterns was made by a man? Correct!

The take-home message here is that people's behaviors in virtual reality are tracked, and therefore can be stored, analyzed, and

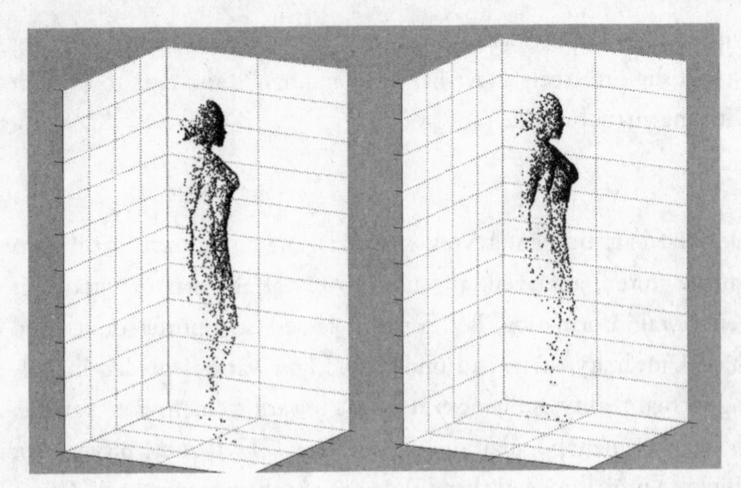

Big Brother is watching: The density of the dots in the graphs depicts where the students' attention was focused.

used—for good, bad, or whatever the person collecting the information wants. In virtual reality, interested parties are able to track and store every micro-movement, gesture, eye gaze, speech, etc., in much the way that the content of e-mail and instant messages can be recorded.

SEEING IS BELIEVING

When people walk a plank over a virtual pit, they often balance themselves by extending their arms out from their sides. These movements can be tracked and then rendered to graphically depict their arms. The tracking data provide precise mathematical specification of the arm movements at any given instant. The computer renders scenes frame by frame. Each time the computer receives the digital arm-tracking data, it creates the underlying structure of the arm in the right location, makes sure the surface image or texture of the hand is appropriate, and takes care of subtle cues such as lighting, shadows, and the occlusion of objects. Such rendering (moving from abstract mathematical data to an image that looks natural) has typically been a thorn in the side of virtual-reality programmers.

Today we take for granted how quickly computers render scenes, but one of the earliest and biggest challenges for virtual-reality programmers was keeping processing speed high enough to render the movements of users in high fidelity. One of the first objects ever rendered in virtual reality was a wireframe cube.

While wandering around the suspended frame of a cube may not be the most exciting virtual experience, Ivan Sutherland, a wizard of computer graphics and longtime professor at Berkeley, was well aware of the tradeoff between how detailed and natural graphic objects appear, on the one hand, and the speed of rendering on the other. A detailed scene takes longer than a plain one for a graphics computer to draw. Consequently, in complex scenes, it's not possible for the scene to keep up with changes in the user's point of view. If one moves toward a physical object, then it grows larger in her visual field. As she moves a millimeter closer to that object, the computer has to render a slightly larger version of the object to her eyes. Another millimeter and it has to do so again, another millimeter and again, etc. Sutherland's selection of the cube—a simple graphic object composed of six polygons—minimized the complexity of the object, thus reducing the time necessary for rendering successive frames.

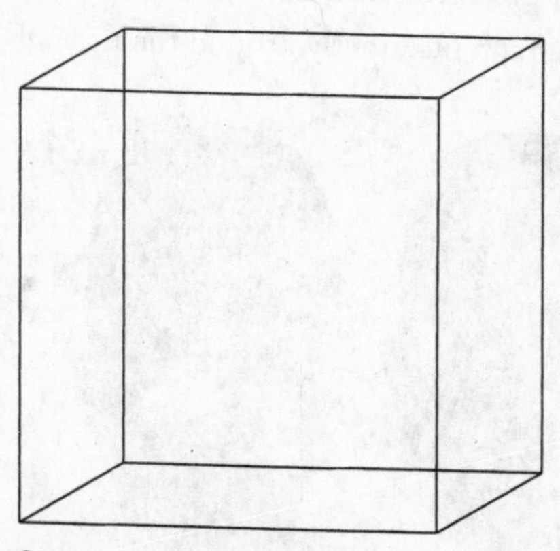

Simple polygons: One of the first objects rendered in virtual reality.

Had the virtual scene at Sutherland's lab in 1968 been more complex—say, a detailed re-creation of the Grand Canyon—it would have taken an impossibly long time to update the scene to accommodate the user's movement. This is called "lag" and can make for unpleasant side effects, which users in *Second Life* even now sometimes suffer. Lag occurs when the system takes too long to render the virtual world. If the system draws updates about sixty times per second or faster, then the updates to a person's viewpoints occur frequently enough to be experienced in a natural way. However, if the graphics are so detailed that it takes a long time to render, when the user turns her head, it takes noticeable time for the proper viewpoint to arrive. When there is a long lag, someone had better be available for cleanup duty, because nausea (i.e., "simulator sickness") can result.

Look at the three renderings of the Stanford undergraduate below.

We can illustrate two important concepts with this figure. One is the differentiation between geometry and texture. The wireframe is the geometry that makes up the underlying structure (seen in the figure as a subtle white mesh). Next, there is a texture, an image of the face, wrapped around that wireframe. So, if one were to wrap a photograph of Brad Pitt around the wireframe, taking extreme care so that the nose from his photograph fits exactly around the nose of

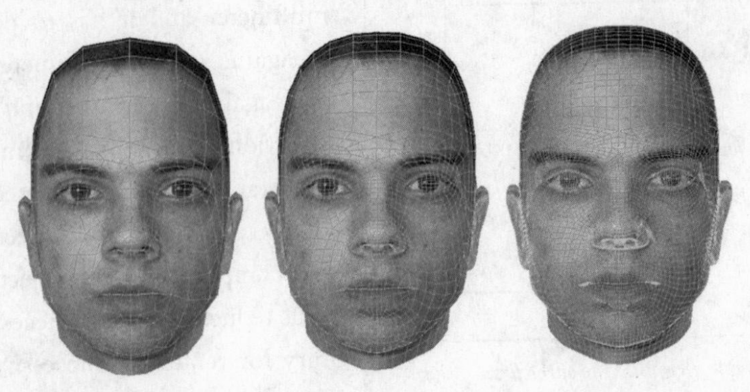

Head models, from rough to smooth.

the wire head, one would have added a layer of texture. Of course, the viewer would not see the mesh inside a finished rendering; she would see only the texture and shape of it.

The second concept is complexity. As you look at the figure from right to left, the structural geometry gets less detailed and chunkier. The one on the left has more little bounded spaces (i.e., polygons) than the one on the right. The advantage of adding more polygons is a smoother rendering that looks more like an actual physical object, in this case a person. The disadvantage of adding more polygons is that it costs processing time. As of this writing, it takes roughly three times as long for a PC to render the head on the left than the one on the right.

The more finely geometrically structured the 3-D models used in a simulation, the higher the probability of a lag and the subsequent discomfort of simulator sickness. With cubes, it doesn't take many polygons to render appropriate structural geometry. Realistic human faces take many more. For a full head of human hairs, it becomes extraordinarily difficult, which is why in animated movies, such as *Shrek*, the animators will often sneak in a technical "show-off scene"—for example, one in which a knight takes off his helmet and shakes his perfectly flowing hair in the wind.

Tracking and rendering make up two-thirds of the virtual-reality processing cycle—movements need to be tracked accurately and often, so that the computer can render the viewpoint at a high speed. The third crucial part of the formula is the display.

DISPLAYS

A display connected to a digital computer gives one a chance to gain familiarity with concepts not realizable in the physical world. It is a looking glass into a mathematical wonderland.

—IVAN SUTHERLAND, 1965

How exactly does a user see and hear what is rendered? First, immersive displays should surround the user—for example, by providing stereoscopic images and sound as well as sufficiently wide field of vision to allow peripheral sight. Accordingly, the experience is a worthy replacement of a physical one, allowing the virtual information to be perceived as a natural scene. Second, stimuli from grounded reality should be blocked out. In other words, if the person in the digitally rendered pit can see the carpet on the physical floor out of the corner of her eye, then the immersive effect of the simulation is diminished. Consequently, the most immersive display devices prevent users from seeing the physical world around them. From the earliest days, the military has been aware of this and took great care to develop flight simulators in which pilots were totally surrounded by the fictional cockpit and airspace.

The earliest displays were built by Ivan Sutherland. One of them is displayed at the Computer History Museum in Mountain View, California. Sutherland's 1968 creation is housed in a glass display case toward the back of the museum.

This model used cathode ray tubes to project the same image to each eye (i.e., "monoscopic"). Today's counterparts allow binocular stereoscopic vision and, hence, the experience of visual depth. Also, instead of displaying colored objects, Sutherland's early device displayed only monochrome wireframe objects, that is, only the bar-

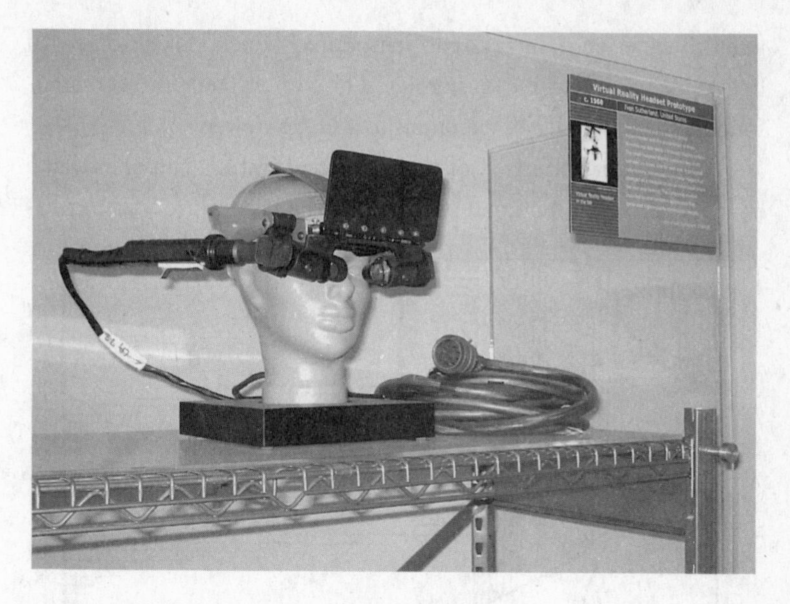

An early prototype of the head-mounted display.

est outline (i.e., structural geometry) of a transparent object without color. In those days, getting computers to chug through software code to render complex images in near real time was impossible, so the objects in virtual worlds were kept simple. Also, Sutherland's helmet was so heavy and intimidating that it had to be tethered to the ceiling to alleviate its weight on the user's head and neck. Indeed, his head-mounted display was nicknamed a "Sword of Damocles." Our colleague Andy Beall, taking a much less weighty approach, created the first head-mounted display at UCSB by successfully rigging two tiny Sony Watchman screens and a scuba mask.

One of the most famous displays ever built, the precursor to Sutherland's "Sword of Damocles," was the famous "Sensorama" developed by Morton Heilig, a filmmaker who believed that movies could better be enjoyed via all five senses rather than just sight and sound. To facilitate this experience, he created "Sensorama," which looked somewhat like a video arcade game on steroids. The console could project sights, sounds, smell, wind, and vibrations to a single viewer.

Heilig built about ten of these devices, and it's rumored that a handful are still functional today. Over the years, Jaron Lanier has

The Sensorama "theater."

used at least half of these. When asked about his favorite virtual experience in the Sensorama, he described "[a] date with a smiling hippie girl." The simulation starts out with a picnic in a park, the sharing of a soda pop, and moves on to a ride in an amusement park, with smell, wind, sights, and sounds. Lanier assured us that it was an innocent date, akin to "puppy love," nothing like the things one can find on the Internet today.

One of the most important aspects of a virtual display is that it allows users to see the world naturally. This requires a large enough *field of view* to see not just what is on the computer screen, as if gazing through a small window, but instead to have a wider view that allows for peripheral vision. Indeed, Thomas Furness, who worked for the U.S. Air Force in the early 1980s, was one of the first scientists to put on a head-mounted display with a wide field of view.

Furness said, "I felt like Alexander Graham Bell, demonstrating the telephone. We had no idea of the full effect of a wide-angle view display. Until then, we had been on the outside, looking at a scene as if we were moving back and forth and looking through a window. Suddenly, it was as if someone reached out and pulled us inside." Today's head-mounted displays allow for a wider view, but even the best ones available today fall short of the human field of view.

There are other types of displays, ones that are not worn but rather surround the user, such as CAVE technology. Stereoscopic visual effects are created in CAVEs by having users wear 3-D glasses coupled with 3-D projection, much like early 3-D movies. The size of the CAVE itself should not be confused with the size of the virtual world that can be displayed—it is possible to create the illusion of depth in very large virtual worlds, such as mountain ranges and oceans.

Still others you've likely experienced yourself. They involve large movie screens, such as IMAX viewed through 3-D glasses. In Dis-

An illustration of CAVE technology.

ney's California Adventure theme park exhibit "California Soaring," visitors sit in seats that are suspended and controlled by hardware that re-creates the rolls, pitches, and yaws of the hang glider. The seat moves in sync with films of various outdoor California scenes that are projected on a very large IMAX screen. Visitors overwhelmingly feel as if they are actually hang-gliding through California.

Some displays are the stuff of science fiction that may or may not eventuate; for example, a device built into a pair of eyeglasses that literally uses the full-color spectrum of light and actually "paints a picture" on the retina. Another is the prospect of tiny monitors built into contact lenses and powered by body heat.

Regardless of the display technology, the basic concept behind current implementations of virtual reality is cyclical. Track the user's point of view, use the tracking data to render the appropriate scene, and send it to the user's display. Today, we are many objects and polygons removed from the simplicity and starkness of wireframe cubes.

We often get asked, "When is virtual reality going to be in every home? When is it going to get huge?" And every year, for decades, we've answered, "Next year." There are constant predictions about when virtual-reality technology will be available everywhere, and they are invariably incorrect. But the movement of people into various types of virtual worlds is increasing at an exponential rate. Virtual worlds displayed on smart phones and computers—for example, *Second Life*—are proliferating. Immersive worlds are also gaining popularity, appearing in more places than ever before. Our prediction is that the emerging tracking and rendering technologies offered by Nintendo's Wii, Microsoft's Kinect, and PlayStation's Move will combine with 3-D monitors and inexpensive head-mounted displays to increase consumer demand substantially in the next few years. In other words, virtual reality will soon become the killer app of the multibillion dollar gaming industry. When this happens, the corresponding influx of funding will greatly accelerate the pace at which virtual reality will enter the common experience.

OF AGENTS AND AVATARS

When Jaron Lanier coined the term *virtual reality*, he added the latter word to mean social worlds with people in it. At that time, other terms described extant "virtual worlds" such as flight simulators, which date back to the first half of the twentieth century. By adding the word *reality*, he intended to label simulators with people interacting within them. Indeed, in the next few years, the most significant uses and implications of virtual reality will push us well past the enjoyment we derive from a Wii game or even the astonishing immersion of virtual-pit worlds. More sophisticated immersive virtual-reality technology allows us to simultaneously track much more, including winks, facial expressions, and shoulder shrugs. This is particularly important because things become much more interesting, and a bit more complicated, when a visitor is not the only person in a virtual world.

Virtual human representations come in two flavors: *avatars*—representations driven (i.e., tracked and rendered) by the actions of actual people, typically in real time; and *agents*—representations driven by computer programs. Neal Stephenson, the author of the immensely popular mid-1990s virtual-reality novel *Snow Crash*, is largely credited with applying the term *avatar* to digital human representations, though there is some debate, as researchers from Lucasfilm created an online world with "avatars" specifically labeled as such almost a decade before *Snow Crash* was published.

Regardless of who first used the term *avatar* in the context of virtual reality, it was originally a religious term. For example, Vishnu was a Hindu god that crossed over to Earth in many different forms, or "avatars." In general, an avatar was typically defined as an incarnation of a deity. The gods of Eastern religions who visited the Earth were embodied by some type of recognizable human or animal form. Similarly, when humans visit digital space, they often are embodied in some way to carry out their actions; Stephenson referred to this form as an "avatar."

Imagine standing in a living room, walking up to a virtual mirror, and seeing an avatar as your reflection. Using tracking technology, we can create a very functional virtual mirror. Users report to us that one of the most striking virtual experiences is to observe one's avatar (with one's own photographically realistic image) in the virtual mirror, and then to watch as it morphs magically into the face of another person. It's not uncommon for someone to exclaim, "What happened to me?!" or to demand, "Change me back!" On the other hand, urgent requests to "please make me that way in the real world" are not uncommon, either. A person can literally see, hear, touch, and even smell herself in a virtual mirror.

Stephenson foresaw how people would use avatars:

Who's the fairest one of all?

> *Your avatar can look any way you want it to, up to the limitations of your equipment. If you're ugly, you can make your avatar beautiful. If you've just gotten out of bed, your avatar can still be wearing beautiful clothes and professionally applied makeup. You can look like a gorilla or a dragon or a giant talking penis in the Metaverse. Spend five minutes walking down the street and you will see all of these.*

Imagine using our mirror simulation so that people experience what it's like to be a member of another race, not unlike the experiences John Howard Griffin recounts in his 1961 best-seller *Black Like Me.* Or imagine a male supervisor spending some time at work as a woman, as opposed to reading a handbook on how to avoid sexual harassment. If a picture is worth a thousand words, a virtual-reality experience is worth thousands of photographs. Indeed, we have been working with corporations to produce virtual-reality tools that will help train people in issues of diversity (more on this concept later on).

Let's consider the social implications of all of this within the context of the ever-improving technology we looked at earlier. The popular online world *Second Life* is currently quite simple. The display is a simple, two-dimensional computer screen. But with eighteen million registered accounts, *Second Life* is popular despite the fact that, as far as virtual reality goes, it's not very immersive perceptually. The reason for its popularity is that it *networks* many people, allowing them to share social experiences. Everyone creates their movements via their own keyboard by hitting keys. *Second Life* automatically transmits these keystroke data—if one person moves three feet forward, *Second Life* then automatically notifies everyone in the virtual vicinity by updating their avatars' position on those

A day in the Second Life.

users' displays. In this sense, inhabitants not only have their own field of view updated as they move around, but they also see others as they move around the virtual space. The current version of *Second Life* will someday seem quaint, similar perhaps to how the original Atari looks to us now. But just like Atari presaged the Xbox, Playstation, and Wii, networked online platforms such as *World of WarCraft* and *Second Life* provide strong hints as to where virtual social interaction is headed.

It's one thing to see your own avatar shift shape and color on command. Doing this alone is largely a novelty. Imagine being able to change your age, gender, weight, height, and even your species at the snap of a finger at a cocktail party or in a business meeting. In a virtual world, many avatars can come together. It's just as easy to render the avatar as the Jolly Green Giant as it is to replicate the spitting image of the user, and anywhere in between. So one reason social virtual worlds are becoming so popular is because of this alluring but potentially dangerous idea of appearing however you want, whenever you want.

CHAPTER FOUR

WINNING VIRTUAL FRIENDS AND INFLUENCING VIRTUAL PEOPLE

As people spend an increasing portion of their time in virtual reality, psychologists, sociologists, anthropologists, and other scientists are afforded a remarkable opportunity to study social relationships and interaction. The amount of data captured represents nothing short of a revolution in the behavioral sciences. In later chapters, we'll turn to how we use, and will use, virtual reality to construct new worlds and possibilities. But for now, let's examine how we're using virtual reality in order to get a better handle on ourselves.

In 2000, Chuck Noland was the lone survivor of a plane crash somewhere in the South Pacific. Stuck for years on a small tropical island, Noland provides a case study in human isolation. Fairly early in his plight, the stranded Noland became so needy of human contact that he created a friend—in virtual-reality terms, an agent—fashioned from a volleyball washed ashore from the plane wreck that stranded him. Using his own blood, Noland painted a primitive face on the ball and named him "Wilson," treating him as a close friend and eventually taking him along on his handmade raft in a desperate attempt to return to civilization. Tragically, during that journey, Wilson was lost during a storm at sea, and Noland was devastated.

We're talking here, of course, about Tom Hanks's performance with his volleyball costar in the movie *Cast Away*. While just a movie, the central drama of the film—complete human isolation—struck a chord with millions of viewers, and indeed cut right to the heart of the fundamental human need for contact. Many moviegoers were so moved by the interplay between Hanks and his little friend that the Wilson Sporting Goods Company actually marketed a "Wilson" volleyball with the "blood"-drawn face.

As gregarious and social creatures, humans find being alone for extended periods, whether by a twist of fate or deliberate ostracism, highly undesirable at best and severe punishment at worst. Purdue

Wilson, Tom Hanks's volleyball costar in Cast Away.

University social psychologist Kipling Williams, a very friendly and outgoing person, is the world's leading authority on social isolation, or ostracism. Williams labels ostracism as "social death." His research reveals that nearly everyone experiences or contributes to some level of ostracism almost every day, and that even minimal ostracism can be painfully distressing. Important, Williams demonstrates that people are so averse to ostracism that they not only seek out other people but sometimes settle for imaginary friends, as in the case of Noland's creation of Mr. Wilson.

Williams reports that many years before he began investigating ostracism,

> *My dog and I were lying on a blanket in a park. A Frisbee rolled up and when I turned to see what it was, I saw two guys looking my way, expectantly. I threw the Frisbee back to them and was about to sit back down when they threw it back to me. I joined them. We didn't speak, but it was fun and I felt welcomed. We threw the disc around for about 2 minutes. Then, just as suddenly as I was included, I was shut out. They stopped throwing to me; they stopped looking at me. It was as though I was suddenly invisible and had never existed. This experience, with strangers, was surprisingly powerful and negative. I felt terrible and awkward and helpless. I slinked back to my dog showering her with praise and affection. When I regained my composure, I realized two things: being excluded from this game of toss upset me, even though it was with strangers, and I found a way to study ostracism in a laboratory.*

Williams went on to create a research scenario to examine the effects of ostracism online, using a simple ball-tossing game he called *Cyberball*. In a typical ball-tossing paradigm, the participant in

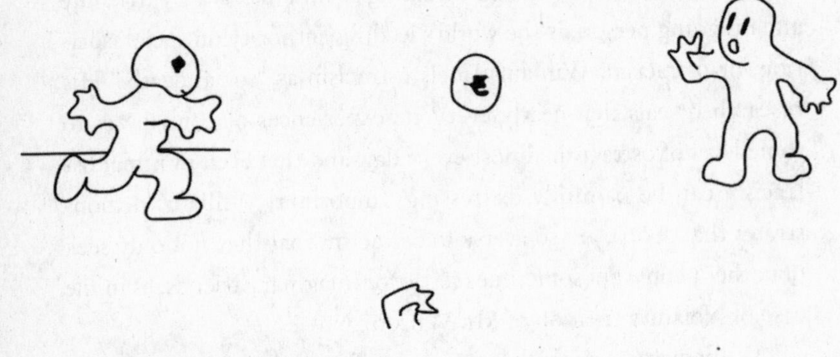

Kip Williams's Cyberball.

the online version is initially included in the virtual ball-tossing but soon is left out.

Imagine playing *Cyberball* in the cramped tube of a magnetic resonance imaging (MRI) machine. You see the online images projected via a small mirror reflecting the scene to your eyes as you are lying inside the tube looking up. Initially, the ball is tossed to your avatar (i.e., hand) and you can toss it via button presses to either of the other supposed players. If you are in the critical experimental condition, you soon find the virtual players hogging the ball, tossing it only between one another. You are excluded from the ball-tossing. Alternatively, if you are in the control condition, you are included in the ball-tossing throughout the game. The results of the MRI study demonstrated that the same areas of the brain that are active during physically caused pain, like having a broken arm or a severe burn, are active during ostracism from a virtual ball-tossing game.

Williams concludes that in order to avoid the pain of ostracism, people will do whatever they can to reconnect with others, even if, like Noland in *Cast Away*, they need to use their imagination. Sadly, others vent their pain and rage psychopathologically, like the social outcasts who perpetrated the Columbine High School massacre by shooting fellow students, or the hermit-like Unabomber, who chose to engage others by blasting them via letter bombs.

COMMUNICATION AND SOCIAL INFLUENCE

Social psychology has long been defined as "the study of how the thoughts, feelings, and behaviors of individuals are influenced by the actual, implied, or imagined presence of others." Indeed, Gordon Allport, born in 1897, proposed this enduring definition and is widely regarded as one of the fathers of modern social psychology because of it. His intuitive conclusion regarding "implied" and "imagined" presence of others provides a perfect framework for modern concepts like "virtual reality," "agent," and "avatar."

To simply illustrate Allport's definition, consider something as mundane as the grooming behavior of an apartment roommate. If your roommate actually had a date waiting in the apartment, his grooming behavior would easily be recognized as an example of being influenced by the *actual* presence of another. If his date was not yet physically present but on the way over, his grooming behavior would appear to be influenced by the *implied* presence of another. If he was prepping just in case he got lucky and found a date, his grooming behavior would appear to be influenced by the *imagined* presence of another.

The term *social influence* refers to the general types of effects that people have on each other. While persuasion is one type, social influence takes many forms, ranging from the way people keep their distance from one another while standing in a group to the way they copy and mimic one another's behavior. Social influence can result from words, intonation, or gestures and facial expressions. It can occur via sight, sound, touch, or even smell.

A common view of biologists and psychologists is that humans are driven to interact with others by evolutionary imperatives, so much so that we acknowledge a "man in the moon" and teach young children to say "Good night, moon" before drifting off to sleep. The physical absence of others does not appear to be very important. Rather, it is the lack of social contact or communication with others that is important, as Williams's work on ostracism aptly demonstrates.

Media developments have made humans ever more social. Historically, advancements in communication media technology have expanded humans' ability to make contact with others, physically or virtually, in ever more ways.

In their landmark book *The Media Equation*, Byron Reeves and Clifford Nass, our colleagues and scholars in the department of communication at Stanford University, argue convincingly that people treat communication technologies as if they are people. Together and separately, Nass and Reeves have been studying the relationship between people and virtual media for more than twenty years, conducting dozens of experiments confirming that people socialize with virtual agents the way they do with other people. In their widely discussed "politeness study," participants who used a computer were more positive about the computer's performance if the relevant evaluative questions were asked "by" the same computer that was being evaluated. In contrast, they were more critical if they evaluated that same computer while answering the evaluation questions on another machine. In other words, people did not want to "offend" the machine while evaluating it!

Reeves and Nass have turned that academic knowledge into dozens of well-known applications. For example, they worked directly with Microsoft to develop Clippy, the virtual agent paper clip, a familiar icon to those who have used Microsoft Word over the years. Clippy is loved by some and loathed by others, but one thing that is difficult for anyone to deny is that this agent evoked strong emotional reactions. Indeed, in 2005, an online group began an Adobe Photoshop contest called "Clippy Must Die," inviting artists to depict the fatal demise of the icon. The reaction was so strong that, in the late 1990s, Clifford Nass himself mentored an honors thesis by Stanford student Luke Swartz, titled "Why People Hate the Paperclip."

The importance of communication for humans cannot be overstated. As every medium from storytelling to virtual reality illustrates, humans are an eager lot when it comes to interacting with virtual beings, and virtual reality is revolutionary because it facilitates highly complex social networks.

Whether the ever-increasing proportion of the human population that is networked via the Internet is, on balance, good or bad for the human species is debatable. More important, virtual reality is a *new* thing, one we must take pains to understand so that the human condition can be improved rather than impoverished. Like storytelling, printed books, the telephone, radio, and television, virtual reality is revolutionizing the human condition, facilitating social interaction with more "others" than our ancestors ever imagined.

SLEIGHT OF MIND: A GENERAL THEORY OF VIRTUAL BEHAVIOR

Over the last decade we developed a general theory of social influence to guide our understanding of people's behavior in virtual reality, and conducted numerous experiments to test and refine

The window-washer statue on State Street.

it. As a result of our efforts, we have identified five important factors that govern the way people engage with others within virtual reality.

Theory of Mind: Our Beliefs About the Sentience of Others

In Santa Barbara, California, there are many small, pedestrian-only shopping areas called *paseos* ("paths set aside for walking"). One *paseo* in particular presents a frequent challenge for shoppers and—to a behavioral scientist's way of thinking, at least—stretches their theory of mind. During our first visit there, while sitting at an outside table in a café, we saw a workman, dressed in white coveralls, work boots, and a blue hat, holding a bucket and using a squeegee to wash a large plate-glass window. His right arm and hand was extended upward, and he was about to run the squeegee down the pane. We gave him no thought except perhaps that he must be an excellent washer, as the windows were exceptionally clean. Yet, after a short while, we began to feel there was something not quite right about him. After a few more minutes, we got it! He was not moving and had been in that same position the whole time we were there. Indeed, he had probably been there twenty years or more. He was not an excellent window-washer at all—"he" was not even human. Our perception of the figure had changed from "excellent human window-washer" to "amusing statue."

People have a penchant to diagnose, or make attributions about, other people's mental states such as beliefs, intentions, attitudes, motivations, knowledge, and personality. He's dangerous. She's flirting with me. He's politically conservative. These conclusions exemplify "theory of mind" and set the stage for how social interactions play out.

As the window-washer statue demonstrates, there is a much more basic assumption at work as well. When people see an approaching human-looking form or hear a human-like voice, they typically assume these "human forms" are sentient and capable of normal human activities. That is, they assume these "forms" are human. Social psychologists call this process making an "attribution of sentience." People use many sources of information to make these assessments.

A first clue is the "animacy" of others' movements, meaning how realistic they appear to be walking, looking about, catching others' gaze, etc. Our window-washer was in a posture that evoked a natural sense of impending movement, but strangely, it never happened, which triggered a red flag in our minds. Indeed, at such a level of perception, people typically are not even conscious of making the attribution of sentience, except, of course, when a human turns out to be a statue or, perhaps, when you realize the rustling outside your bedroom window is a tree branch, not a burglar. In the physical world, verifying sentience is often quite straightforward. However, virtual reality presents a different challenge, because people are not physically face-to-face with one another while they interact. What is most important in virtual reality is the *belief* that a person holds about a human representation (for example, is it really an actual person—an avatar, or just a human-appearing computer algorithm—an agent).

In general, if a person believes that a representation in virtual reality is an avatar, she will make an attribution of sentience and interact more or less as she would in a physical, face-to-face situation. On the other hand, if a person believes that the representation is an agent, the attribution of sentience is not as likely. It should be noted that theory-of-mind beliefs are also important in the physical world, where people sometimes treat nonsentient but live human bodies as though they are sentient (Terry Schiavo in a "persistent vegetative state" comes to mind), or sentient and live bodies as though they are not. For example, the famous sociologist Erving Goffman pointed out long ago that people often treat others as if they are not there, or, in his words, "as furniture." For example, when people are in a very private conversation, they often speak loudly enough to be overheard by others in the vicinity—building custodians, taxi drivers, etc., which is why such occupations are often used by spies.

When interacting with a person in virtual reality, theory of mind is not enough. Many other important factors come into play. Even agents can influence us if they have the right moves.

Communicative Realism: The Poetry of Motion

It's an old joke that with his hands tied behind his back, a man from Southern Italy would be unable to speak. The use of gesture and nonverbal behavior is so intrinsic to his production of language that he cannot function without it. Of course, in many cultures, nonverbal behavior is more subtle—for example, in Japan. There is a large body of literature assessing just how important nonverbal expression is for communication and social interaction. Judee Burgoon, a scientist at the University of Arizona who has been studying nonverbal communication for more than thirty years, estimates that over two-thirds of communication among people occurs nonverbally.

In virtual reality, nonverbal behavior is particularly relevant. Nonverbal communication is largely a function of its three component variables. These variables are: movement realism (postures, gestures, and facial expressions, for example), anthropometric realism (recognizable quality of human body parts), and photographic realism (how much these representations look like an actual human).

The first, movement realism, is how well virtual body parts move. Nearly all social transmissions require movement: for example, waving a hand back and forth to signal hello, or vibrating the vocal cords to produce speech. When movements are realistic, people are more likely to be influenced by a virtual representation. The second, anthropometric realism, is the presence of body parts that typically are used for communication. As a virtual representation comes to resemble an actual human, its potential to communicate increases exponentially: once an avatar has eyes, a mouth, and hands, it then can wink, scowl, give a thumbs-up, etc. Third, and less important than the other two, is photographic realism. For example, a virtual representation depicted in high definition may look good during a sporting event, but these extra pixels are merely cosmetic and don't contribute much to social influence.

If doubtful about these arguments, one needs only to watch cartoons. Characters such as Donald Duck, Mickey Mouse, Homer Simpson, Eric Cartman, or Buzz Lightyear demonstrate that movement realism is critically important, along with appropriate support from anthropometric realism (though not all appendages are necessary; eyes, lips, or even just teeth are sometimes enough). Photorealism lags behind in terms of necessity. In other words, even though Bugs Bunny doesn't look like any identifiable human, or even like a real rabbit, the communicative cues we receive from his and other characters' cartoonish ears, eyebrows, and whiskers create a powerful sense of human social agency.

In terms of virtual humans, people commonly confuse how "real" they are with how photographically faithful they are to the physical world. However, these same people would be able to recognize that the images shown opposite depict the same person, President George W. Bush.

In 2007, we conducted an analysis in which we examined every study that had ever manipulated the photographic realism of a face (e.g., "high" or "low" photorealism) to determine the effect of photorealism on social influence in virtual reality. By reading all published papers on the topic, as well as calling scientists and politely asking them for their original datasets, we put together a master database that included approximately eighty experiments from recent decades. By analyzing these data, we were able to determine the exact "effect size" of photographic realism. Unsurprisingly, when a virtual representation has a face, as opposed to merely a simple representation such as a shape with no facial features—or, alternatively, as opposed to just a voice—then it is more effective at influencing people. However, it also turns out that when the face becomes more photographically realistic—for example, moving from the left to the right, in the picture of George W. Bush—there is no increase in persuasion or influence. A simple line drawing is just as effective as a high-definition photograph in terms of persuasion, teaching, and being liked.

In fact, there is even evidence that photographic realism can backfire. Interestingly, as photographic realism of human-like

Most people would recognize that both images depict George W. Bush.

representations, whether robots or virtual agents, nears perfection, people are somewhat repulsed by them. Humans like realism to a point, but there is a specific place at which photorealism that is high but not yet perfect is counterproductive. For example, we all loved R2D2, the *Star Wars* droid, but we often find wax museums with very photorealistic wax models of famous individuals to be creepy and unsettling. Computer scientists labeled this counterintuitive phenomenon the "uncanny valley" as early as 1970, and the phenomenon still receives ample support from scientists and graphic designers today.

With high photorealism and perfect communicative realism, of course, nothing is uncanny and there is no repulsion. However, communicative realism is very difficult to achieve, from a technical standpoint. Consequently, in some animated feature movies, directors are said to have "dumbed down" the photographic realism of their characters in order to avoid engendering uncanny feelings by viewers. In fact, one of the reasons Pixar is so suc-

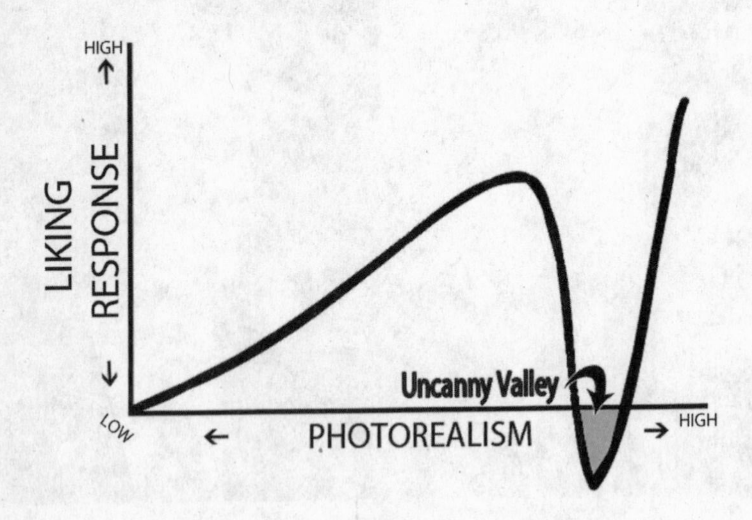

The "uncanny valley."

cessful with their virtual characters, for example Buzz Lightyear in *Toy Story*, is that they do not push photographic realism too far. James Cameron apparently "got it" as well, designing the alien "Na'vi" in his movie *Avatar* to be anthropometrically similar to humans—though much taller, with tails, and blue—but to look different from humans photographically. In this way, the audience was more comfortable with the computer-generated facial expressions and body movements.

If the only two factors necessary to develop a scientific theory to causally explain social interaction in virtual reality were theory of mind and communicative realism, our work on social influence would be relatively straightforward. However, there are at least three other important factors that must be taken into account.

Response System: Unconscious Versus Conscious Levels

Since Sigmund Freud, psychologists have had sort of a love/hate relationship with the idea of unconsciously driven behaviors. Although the notion of the unconscious fell out of favor in behavioral science during most of the early- and mid-twentieth century, the notion of the unconscious stormed back onto the psychological scene with the rise of a more mentalistic, cognitive psychology that emerged in the 1960s. Relatively recently, Malcolm Gladwell's book *Blink* illustrated the effects of these unconscious processes in detail.

If a person believes that a representation is really just an agent, and that agent surprises him by throwing a virtual punch, the person will exhibit a "startle" response of the same magnitude as when a punch is thrown by an avatar. This happens because startle responses are unconscious and, consequently, we don't stop to think that a virtual punch is harmless. On the other hand, if the response was more complex than a simple "startle," then agents might not be so successful at influence. For example, it is much easier to fall in love with an avatar (which happens with surprising frequency online in places like *Second Life*, resulting in many subsequent marriages in the physical world) than an agent (not too many people have expressed a desire to spend personal time with Microsoft's Clippy).

Self-relevance: What's in It for Me?

Some social scenarios are highly self-relevant—in other words, they are extremely important to the people involved. One can hardly think of a more highly self-relevant social scenario, at least in Western society, than falling in love. On the other hand, mere transactions like buying a soft drink are much less important. Self-relevance can be thought of as a "gain factor" or volume control. Typically, there is not a lot of self-relevance during interactions with strangers in elevators. Asking someone to punch a button is not a big deal. Thus it is not surprising that elevator operators were replaced long ago by robotic agents in the form of elevator button boxes. Similarly, there is not much self-relevance in paying the toll on a toll road. We treat actual people collecting tolls not much differently from the electronic device that does the same job. On the other hand, there is a lot of self-relevance when asking someone for a loan; consequently, the way we interact with potential lenders is fundamentally different. It is important that the degree of relevance of a virtual social interaction to the self varies from person to person as a function of their past experiences, personality, and temperaments.

Context: Where Am I?

People's behavior often depends on where they are. Talking in libraries is different in terms of volume and word choice than at a boxing match. Likewise, the type of communicative realism that contributes to social influence depends on the nature of the specific virtual setting.

When one plays a video game, he intentionally enters the context of fantasy. In this case, anthropometric and photographic realism can be low and players will still become quite immersed. Many games, like *Pac-Man* and *Super Mario Brothers*, became popular despite low levels of communicative realism. On the other hand, a context such as online dating brings a different set of norms and expectations. A caricature drawing likely will not suffice in a profile on Match.com—people want high photorealism in this context.

VIRTUAL REALITY PRESENTS CHALLENGES TO PEOPLE WHO CREATE the simulations as well as people who spend time using them. The solution to those challenges requires basic knowledge about how people socially interact with one another. We have described our scientific theory of social influence within virtual environments incorporating what we consider the most important factors: theory of mind, communicative realism, response system level, self-relevance, and context.

We are not suggesting that this model is the be-all and end-all of such theory, but we believe it is a good start. So far, this theory has allowed us to make substantial progress in understanding how to create effective virtual environments for social interaction and we hope it stimulates others to build upon or revise it as new scientific data warrants.

CHAPTER FIVE

THE VIRTUAL LABORATORY

WHEN THE TWO OF US INITIALLY BECAME COLLEAGUES AT THE Research Center for Virtual Environments and Behavior at the University of California, Santa Barbara, in the late 1990s, one of the tasks we faced was to prove the value of using virtual reality to perform social psychology experiments. The vast majority of our academic colleagues throughout the world had never entertained the idea. Today, everyone knows the word *avatar*. The rise of video games, movies like *Avatar*, and online social worlds like *Second Life* have brought the notion of virtual reality to dinner-table conversations. Many today are concerned about how virtual reality is affecting life as we know it. However, in 1997, it was a different story, and only the most geeky of science-fiction followers were familiar with such terms.

Psychology is like most academic fields—drastic changes in thinking are typically regarded with caution, if not disdain. We could safely assume that many protectors of the status quo would judge our work and decide whether it got funded or published. So, we undertook investigations of well-established social psychological phenomena using virtual reality. We reasoned that if we could replicate classic and well-established social-influence effects using virtual reality, our peers would clearly see its value, accepting our papers for publications and funding our grant proposals.

In this chapter, we describe a number of critical experiments demonstrating that the mind largely treats virtual people just like physical ones. By measuring how people talk, how they emote, how their physiology changes, and how they make decisions, we can demonstrate that virtual behavior is, in fact, "real."

THE STIGMA STUDY

Many people have been stigmatized by others at some point in their lives. Certain social-category memberships can result in stigma, such as being poor, overweight, or a minority. Social stigma results in being devalued by others. We know from several neurophysiological studies that when a person interacts, even cooperatively, with a "stigmatized other," such as a person with a facial birthmark, their brain evokes a pattern of cardiovascular responses that indicates they are threatened. We reasoned that, if people were "really into" virtual reality, they would exhibit the same cardiovascular threat pattern while interacting with stigmatized avatars, and designed an experiment to test this hypothesis.

During the study, participants met another "participant" physically when they showed up at our lab. The other participant was actually a confederate (we'll call her Sally) in cahoots with us. Sally either bore, *physically*, a "port-wine" facial birthmark or did not.

After meeting, the participant and Sally entered virtual reality.

They sat at a virtual table and played a cooperative word-finding game. Our experiment manipulated whether Sally's avatar bore the facial birthmark in the virtual world independently of whether she bore it in the physical world. So:

1. Some participants met Sally with a birthmark on her face physically, and played the game with her avatar that also bore one.
2. Some participants met Sally with the birthmark, and played the game with her avatar that did not have one.
3. Some participants met Sally with no birthmark physically, and played the game with her avatar that also did not have one.
4. Finally, some participants met Sally with no birthmark, and played the game with her avatar that did have one.

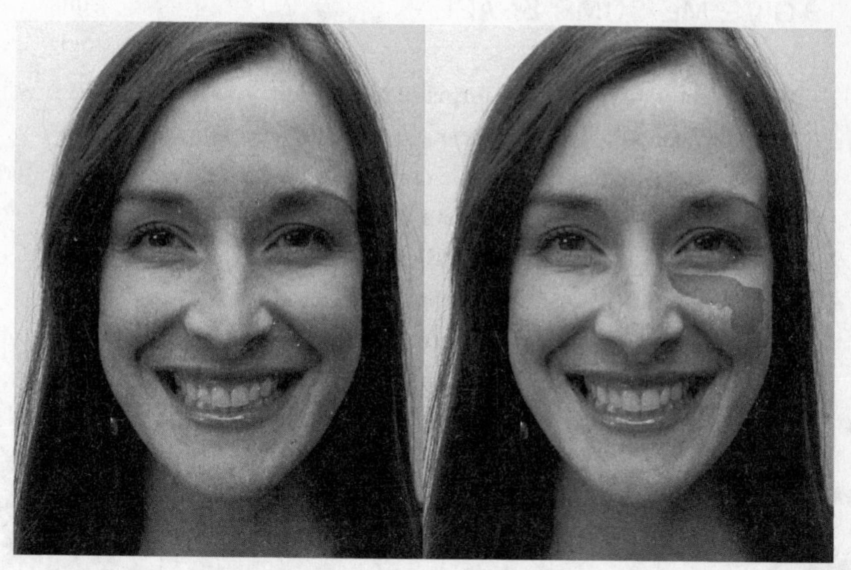

Sally and her avatar.

During the first minute of word-finding play in virtual reality, participants were threatened (as indicated by measuring increases in their heart rates and blood flow, as well as how much their arteries constricted) only if Sally had borne the birthmark outside of virtual reality. So, at the beginning, they were not immersed in a particularly meaningful way. However, by the fourth minute of the task, participants were threatened only if the *avatar* bore the birthmark in the *virtual world*. As one might expect, at first the birthmark from the physical world took precedence. However, just a few minutes later, the virtual birthmark took precedence. Participants knew that the person driving the avatar did not have a birthmark physically but they were still threatened by the virtual birthmark. Just as with prism glasses, people adapt to virtual reality and respond accordingly.

GIVE ME SOME SPACE

One of the most primitive forms of social influence involves what anthropologist E. T. Hall labeled "proxemics." Proxemics refers to the physical distance that people maintain among themselves. Over time, people learn the appropriate interpersonal distance norms, or "rules," which vary strikingly across cultures. Since Hall's pioneering work, hundreds of experiments have quantified the size, shape, and changes in personal space patterns in various social situations.

For example, the interpersonal distance rules in American culture tend to involve greater distances than those in other countries. The United States is a country that likes its space. Indeed, many of its citizens become uncomfortable when a stranger from another culture comes too near. Hall laid out an interpersonal distance schema for American culture, supported by ample research. Simply put, whenever they can, Americans maintain *public distance* of ten feet or more between themselves and strangers, *social distance* of four to ten feet between themselves and casual acquaintances such as strangers asking for directions, *personal distance* of eighteen inches to four feet between themselves and friends, and *intimate distance* of zero to eighteen inches between themselves and intimate others.

Public distance increases when we encounter a stranger who is obviously acting in a culturally deviant way. People give a very wide berth to strangers talking loudly to no one in particular, those walking around naked in public, or others who are socially stigmatized in some way, such as the homeless. Interestingly, cultural proxemic "rules" or norms do not seem to operate when people encounter inanimate human forms. We have fewer problems walking close to mannequins or statues unless the statues are actual humans who suddenly "come to life."

These rules are understandable enough in grounded reality. But we wondered how "personal space" operates between virtual "bodies" in virtual reality, and predicted people would respect the personal space of virtual humans in a similar manner to how they treat physical humans. We devised a simple study to test this idea. Participants found themselves in a virtual room in which a succession of virtual agents stood approximately twelve feet away. Participants were told that their task was simply to approach the agents, one at a time, and learn and remember both the name on the front of the jersey that each was wearing and the number on its back. We told participants that the human representations were agents controlled via computer software. We varied the animacy of the agent in five steps ranging from complete inanimacy—no movements whatsoever and eyes closed—to rich animacy—small, naturalistic body movements and head-turn and eye-gaze toward the participant as he or she approached. We chose the name-number memorization task to motivate participants to approach the agents relatively closely—approximately three feet—and carefully chose the size of the text/number font to ensure that the participants had to potentially intrude into the personal space of the agent to perform their task.

To test how close participants came to the agents, we measured where they walked during the virtual interaction. The results indicated that, indeed, participants walked closer to less animate agents than to more animate agents. This effect was particularly strong for female participants. The gender difference was not surprising to us, because decades of research on face-to-face interaction has demonstrated that females have greater sensitivity to nonverbal information than males.

Participants knew the digital representations were agents merely controlled by computers, but still respected their personal space. This automatic regulation of distance provided us with more evidence to support the notion that people respond to virtual humans in very real ways.

THE HIDDEN PRESSURES OF CONFORMITY

If you have ever walked behind a group of teenage boys or girls at a shopping mall, you undoubtedly noticed the similarity of their clothes, their hairstyles, their lingo, and even the way they walk. That's conformity at work. People conform in many ways to the behaviors and accoutrements of their peers as well. Goths, for instance, have as rigid a style as any sorority sister or fraternity brother.

We decided to see if we could induce conformity effects within virtual reality. Forty years ago or so, in a field study in Reno, we observed that blackjack players tended to raise or lower their bets as a function of how much money other players at their table were betting. We watched gamblers at Harold's Club, a long-since torn-down casino. We randomly chose gamblers playing blackjack alone at a table and we recorded the amounts they bet on each hand dealt to them, which allowed us to compute their average bets while they played alone. When at least two other gamblers joined their blackjack table to play, we recorded all the players' bets by watching them through one-way mirrors, which back then were ubiquitous in casinos.

The results of our naturalistic observational study confirmed that, when joined by other gamblers, the previously solitary blackjack players changed their bets to conform to the average bets of the new players. If the other players bet high or low, then the initial players raised or lowered their bets, respectively, too. A short while after the study run at Harold's, we actually built a laboratory casino complete with "shill" (confederate) players and demonstrated the same conformity effects as we had observed in the downtown casino.

Flashing forward three decades—we began to examine conformity processes in virtual reality. Kim Swinth, a former graduate student at UCSB, eagerly agreed to build a virtual casino complete with an actual blackjack game.

The virtual-reality study we conducted proceeded similarly to the ones we had run three decades earlier, except that we carefully manipulated whether the participant thought the two other players were agents or avatars. In the avatar condition, we introduced the participants to two actors and showed all three of them what their avatars would look like. In the agent condition, we told par-

Place your bets: A participant in our virtual casino.

ticipants that the other two players were controlled by a computer algorithm.

Actually, regardless of which experimental condition the participants were in, the other players were always agents betting according to prearranged plans. The participants played twenty hands of blackjack alone and then were joined by the two other "players" who bet on average either 80 percent more, 80 percent less, or about the same as the participants' average bets at the end of their individual play.

The betting data demonstrate that participants' bets conformed to the betting patterns of the virtual players whether they thought them to be agents or avatars! Again, like the results of the proxemics studies, we were surprised to learn that human gamblers would conform to the bets of other members of a group even when they knew the other players to be agents.

The same conformity patterns occurred in the physical casinos as in the laboratory and virtual casinos. Another of our studies revealed that the virtual currency didn't matter. Participants treated virtual money as real. In a similar fashion, virtual entrepreneurs have an elaborate economy in *Second Life*, in the same way that U.S. dollars were once, in theory, a virtual placeholder for gold.

SOCIAL INHIBITION

The earliest experiment reported in the history of social psychology was published in *The American Journal of Psychology* by Indiana University's Norman Triplett in 1898. His article described both an observational study and an experiment. Triplett's observational study was one of competitive cyclists. Near the turn of the twentieth century, bicycle racing was the NASCAR of its time. It was wildly popular. Triplett noticed that cyclists almost always had faster times when they raced against other cyclists than when they raced against the clock. He wondered if people were generally faster when they performed with other people.

Triplett's experiment looked at children as they reeled in fishing line, either alone or while another child was performing the same task. He reported that half of the children spun faster in the presence of other performers. He also reported that about one-quarter of the children spun slower in the presence of other performers and one-quarter spun about the same. Hence, the effect was not universal, and indeed the presence of others actually slowed down or impaired the performance of a substantial minority of the performers.

What Triplett discovered has come to be known among psychologists as social facilitation and social inhibition effects. After it became well established that the same effects occurred if a person performed solo but in front of a watching crowd, the phenomenon became known as an audience effect.

When one thinks about performing alone or in front of an audience, two scenarios come to mind. An audience can be motivating in a positive (or "performance-facilitating") way, or an audience can exert a negative (or "performance-inhibiting") way. The psychological research on this topic shows that whether the audience is facilitating or inhibiting depends on the performer's expertise. If a person is performing a task with which she is at ease and has performed well in the past, facilitation is likely to be the case. But, if the task is novel or the person has performed poorly on it in the past, inhibition is likely. We've attempted a similar examination with audience effects in virtual reality. Crystal Hoyt, then a graduate student in our lab and now a professor at the University of Richmond, took the lead in developing such a study.

We randomly assigned participants to learn one of two tasks in virtual reality. Both were sorting tasks, in which participants had to categorize numbers and letters. They learned the categorization rules by making guesses and then getting feedback as to whether their guesses were correct.

The participants were randomly assigned to one of three performance conditions. They performed either with nobody watching, or with two agents watching, or with two avatars watching. We also varied which task they performed, either the one that was well-learned or the one that was new to them.

The results indicated that whether an audience was present made no difference when participants performed their already well-learned task—that is, the "easy" one. Furthermore, when they performed the novel task—that is, the "hard task"—participants performed equally well in both the alone and agent conditions: at about 80 percent correct. However, when performing the novel task, participants in the avatar condition performed significantly worse (only about 60 percent correct) than in the other two. This was a classic social-inhibition effect, much like the 25 percent of Triplett's participants who performed more poorly in the presence of others. We believe that avatars inhibited performance more than agents because the latter were not regarded as evaluative—that is capable of judging the participants—while avatars were.

MIMICRY

Adults enthralled with young infants often play mimicry games with them. They open their mouths to see if a three-month-old will mimic them, and are thrilled if they do. They get more complex and stick out their tongues. The appearance of the infant's tongue delights them. When they are a bit older and eating delicious (at least to them) baby food with a spoon, parents open their mouths in anticipation of the babies opening theirs. But mimicry is not always adult-to-infant. It is often also infant-to-adult, as indicated by the fact that very few adults will fail to reciprocate an infant's smile.

Mimicry is regarded by developmental psychologists as essential to the acquisition of communication skills, including nonverbal ones. Mimicry also determines one's "likeability." Imagine going into a job interview and utilizing a special strategy to impress the interviewer in an attempt to get the job. In an experiment that launched dozens of follow-up studies across the globe, Tanya Chartrand, a social psychologist now at Duke University, demonstrated that research participants tasked with performing an interview liked actors applying for the job who mimicked their own gestures more than they liked actors who did not mimic. Furthermore, the participants were consciously unaware of the mimicry. Chartrand labeled this the *chameleon effect.*

We wondered if the chameleon effect applied to digital agents. In the study that resulted, Stanford undergraduates listened to a virtual agent deliver a message advocating a campus security policy that would require all students to carry their identification with them at all times while on campus. We varied the mimicry behavior of the agent, using the tracking data described in chapter 3. In one condition, the agent mirrored the head movements of the participant but with a delay. In the other condition, the agent did not. The participants rated mimicking agents as more persuasive than the other agents, but also rated them as more credible, trustworthy, and intelligent! The effects that occurred were entirely due to the relationship between head movements of the participants and the head movements of the agents; that is, whether they were mimicked movements or not. Is this type of social influence important? Think about how a virtual car sales agent might be programmed to mimic potential car buyers' movements in a digital show room (we will return to this topic later in much more detail in chapter 8).

SO WHAT?

The results of these and other virtual-reality studies demonstrate that virtual behavior is, in fact, "real." In so many facets of social behavior, ranging from financial decisions and the way blood flows through the body, to the manner people stand in a room, people use the same template they use in grounded reality and apply it to agents and avatars in virtual reality.

CHAPTER SIX

WHO AM I?

BY NOW WE'VE LEARNED HOW VIRTUAL REALITY WORKS. THIS PART OF the book explores the next step: exactly how virtual reality and its capacities change *us* as we cross the digital divide, including what we bring back. The changes go right to the core of our individual identities.

From toddlers to seniors, people face the question, "Who are you?" Toddlers will often smile at the question. If they can speak, they might answer "Beth" or "Meridith" or "Greg." If they can't yet speak, they might shake their heads from side to side until they hear their name and then shake their heads up and down. Accidentally dialing a wrong telephone number and asking for a friend usually gets a "Who is this?" reply. People often ask the same question of telemarketers. Not surprisingly, the pat answer to the question "Who are you?" is one of personal identity: "I'm Jeremy," or "I'm Jim."

Virtual agents can also tell you who they are. At the time of its release, audience members were all a bit startled by Stanley Kubrick's *2001: A Space Odyssey* when the spaceship's computer spoke quite naturally, "Hi Dave, this is Hal." Four decades later, audience members were pleased when a little robot was able to say, "Wall-e," in Pixar's motion picture of the same name. Like Hal, digital agents usually can self-identify, such as "Big Blue"—IBM's famous digital chess master—or their "Watson," who can play *Jeopardy* quite well. They can be masters at their games, but at conversation not so much.

More interesting, Ipke Wachsmuth and colleagues at the University of Bielefeld in Germany have created a conversational digital agent, "Max," who, since 2004, has been a big hit at the world's largest computer museum in Germany. Max interacts with museum visitors daily. He has been interviewed by radio and television personalities and spoken with numerous scholars, including us. More important, Max has spoken with thousands upon thousands of museum visitors and continues to greet them daily.

Max tries to be an engaging conversation partner and reveals an array of emotions appropriate to the conversation. Max exhibits emotion not only vocally via inflections and tone but also via facial expressions, breathing, and eye-blinking. If a visitor insults him, Max leaves the scene for a while to avoid escalating a rude interaction. Max also remembers the people with whom he has conversed, storing data such as their name and attributions he has made of their personality characteristics. Like Santa Claus, Max remembers if you've been naughty or nice!

Although we know how virtual agents gain identities—their creators simply name them via software—how people identify themselves is not so simple. Like agents, people are named by their "creators," their parents. But that's not the whole story.

The question "Who am I?" has perplexed psychologists, philosophers, criminologists, and most humans for millennia. What is the core of a person's being? Certainly, it is something more than a

Max interacts with museum visitors.

name. Is there an unchanging core that individuates a person? Is it genetic material or the sum of remembered experiences? Is a person defined by how others view her, or by how she views herself? Is it all of these things and more? There are no simple answers to any of these questions. And virtual reality raises more. Is the person behind an avatar in *Second Life* the same person as the avatar? Most people wouldn't bet on it. Can an agent, even Max, really be more than a name?

Before describing how virtual worlds impact social identity and the self, it is helpful to understand how psychologists conceive of the way people think about themselves. Two psychological concepts are quite important. The first is *social identity*. The second is the *self*.

Individuals play multiple social roles and, hence, have multiple social identities. Many of these roles overlap. *Familial roles* might include son, daughter, brother, sister, mother, father, grandmother, grandfather, wife, and husband. *Social roles* might include friend, confidant, and lover. *Occupational roles* may be professor, engineer, student, therapist, bellhop, and bartender. *Recreational roles* include poker player, golfer, triathlete, and bird-watcher. There are also *group roles,* including race, citizenship, family, club, etc. It is not surprising that if people take stock, they typically find they have multiple identities across many situations. Certainly, some, such as familial, occupational, and group roles, are more important than others. Many overlap, such as "Canadian hockey player."

What ties all of a person's roles together is the self. As mentioned above, what the self consists of is debatable. It includes characteristics that endure over time, such as temperament, personality traits, and attitudes. Scholars believe that some are genetic and others are learned over time, though they are moderated by situations. For example, a person may be quite aggressive on the hockey rink, but quite gentle with her children.

In this chapter, we focus on self-perception; that is, how people interact with and view themselves. People's self concepts are not only influenced by others with whom they interact, but also by their *own* appearance and behavior.

THE SELF-PERCEPTION EFFECT

There is more truth to the lyrics of the famous song "When You're Smiling" than people realize. Although humans smile when they are happy and frown when they are unhappy, the reverse is also true. Smiling can actually cause happiness and frowning unhappiness. Consider psychologists' "facial feedback hypothesis," which posits that facial movement can influence emotional experience. A landmark study conducted by Fritz Strack and colleagues confirmed this hypothesis. These researchers misled experimental participants into believing they were taking part in a study about empathy—more specifically, understanding how difficult it is to accomplish certain tasks without using one's hands.

The experimenters asked participants to hold a pen in their mouths in one of two ways. One, illustrated in the left panel of the figure, is the "lip position," in which the pen is held in pursed lips that contract the facial muscles associated with a frown (pursed lips and a tightened area between the eyebrows). The other, illustrated in the right panel of the figure, is the "teeth position." Here holding the pen affects the facial muscles associated with a smile (open lips with the outer corners of the mouth drawn upward). And, of course, as in most experiments, there was also a control group that held the pens not in their mouths but in their nondominant hands.

While holding the pen in one of the three ways, the participants rated how difficult it was to write while holding the pen. But the target task, which participants did not know was the actual objective of the study, was rating how funny they thought a cartoon was. As predicted, participants in the "teeth," or smile, condition rated the cartoon funnier than those in the "lips," or frown, condition. By independently manipulating the appropriate smile or frown muscles, the experimenters changed how funny people thought the cartoon was!

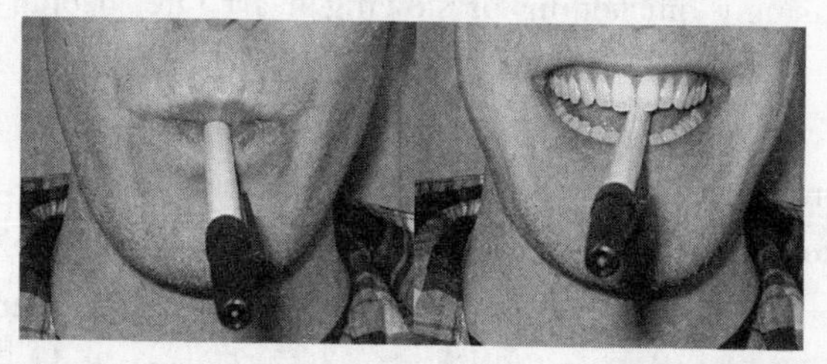

Activating smile and frown muscles: "lip position" and "teeth position."

Social psychologist Daryl Bem actually began graduate work in physics at M.I.T. but became so intrigued with the civil rights movement that he decided to switch fields to pursue a career in social psychology. He developed "self-perception theory," which maintains that when interacting with others, people unconsciously or consciously clarify notions about themselves by observing their own appearance and behaviors. Dozens of studies in the past fifty years have supported Bem's theory by demonstrating that people are constantly influenced by their own bodily feelings, appearance, and utterances as they make all kinds of decisions in daily life.

During a groundbreaking self-perception study in the 1960s, conducted by Stuart Valins at the State University of New York at Stony Brook, male participants heard "beeps" corresponding to their heartbeats while they viewed *Playboy* centerfold models (sorry, no figure here). Not surprisingly, the beeps were normal (about seventy-two beats per minute) during the viewing of some photographs, but tweaked by the experimenters to indicate a faster or slower heart rate during others. Afterward, participants rated the attractiveness of all of the models. Valins found that participants rated the models whose pictures were accompanied by faster or slower heart-rate feedback as more attractive than those accompanied by normal heart-rate feedback. In other words, when Valins "faked" an increasingly quickening or slowing heart rate, people actually reported the models to be more attractive than when they heard beeps at a normal heart rate. At some level, participants assumed that photographs of some attractive centerfolds caused changes in their physiological state, while photographs of others didn't. When participants observed changes in their heart rate, they inferred that it must have been due to the attractiveness of a centerfold. Thus, an observation of their *own* responses led participants to infer their liking for the centerfolds.

Not only do bodily feelings cause people to reassess themselves, but physical appearance does as well. Think about something as simple as the color of clothing. Penalty data from the National Football League demonstrate that the Oakland Raiders, reputedly the league's "thugs," play more aggressively when wearing black, as distinct from white, jerseys. Some might take issue with cause and effect—perhaps Raider transgressions were simply more visible because of the color, or perhaps the Raiders are more aggressive because they're generally wearing their black jerseys when they play at home, with a crowd to fire them up. However, Thomas Gilovich at Cornell University used a very creative study to get to the bottom of the "Raiders Syndrome." His team analyzed past records from the National Football League and the National Hockey League to show that teams wearing black uniforms received more penalties than teams wearing uniforms of other colors. Given that teams often change uniform colors depending on the location of play, Gilovich was also able to show these penalty differences even within the same teams, when they switched uniforms.

Of course, this does not rule out the possibility that referees are more likely to perceive an athlete in a black uniform as being more aggressive. Consequently, the researchers turned to a laboratory study to see if they could actually cause more aggressive behavior via apparel. In this study, experimental participants came to the lab and dressed in either black or white uniforms. Once dressed, participants chose five games from a large list of games in which they would like to compete. The games varied in aggressiveness. Results indicated that participants in black selected more aggressive games (e.g., "chicken fights," "dart gun duel," and "burnout") than people in white (e.g., "basket shooting," "block stacking," and "putting contest"). Participants were unaware that what they wore had anything to do with how they behaved, so the causal process was unconscious.

As Gilovich argues, "[J]ust as observers see those in black uniforms as tough, mean, and aggressive, so too does the person wearing that uniform." Hence, observing their own appearance (i.e., "I am wearing a black uniform") led participants to infer their disposition (i.e., "I am an aggressive person"), which, in turn, led to changes in behavior (i.e., "I will select more aggressive games").

The ante is upped when appearances are modified to have a far more explicit connotation. Researchers conducted an experiment in the late 1970s, in which participants were randomly dressed as either nurses or Ku Klux Klan members. Participants were asked to deliver electric shocks to help a failing person learn a particular task by doling out punishment for incorrect answers. The study revealed that those who were arbitrarily assigned to wear a nurse's uniform were more compassionate and delivered fewer shocks.

In sum, in the physical world, many of our actions are actually driven by seemingly arbitrary surface features—for example, facial expressions or the color and style of our clothes. Personal characteristics, such as humor and aggression, can be changed by a dark shirt or a particular facial configuration, though we are generally not conscious of this process.

THE PROTEUS EFFECT

In their day-to-day lives, people often consciously manage their appearance as they dress in the morning. Men shave, women put on makeup, and business people don "power suits." People also diet, exercise, and tan. Some individuals take self-presentation to a more involved level, injecting Botox, enduring plastic surgery, or switching their genders via a combination of hormones and surgery.

Virtual reality, however, can take strategic self-presentation to another level. Unconstrained by physical reality, everything about a person's appearance and behavior is up for grabs. One can be effortlessly transformed by "donning" an avatar, which can be taller, the other gender, or even a different species than the person wearing it. And people alter avatars with great frequency. Like the Greek god Proteus, avatars can take on whatever form the person in control desires.

In 1995, Sherry Turkle, a scholar of culture and technology at M.I.T., published her famous book, *Life on the Screen*. It was based on Turkle's extensive experience observing people socially interacting within multiuser domains, or MUDs, which, at the time, were text-based online worlds where people were free to create their own identities. In this pioneering work, Turkle not only documented some of the first uses of role-playing via digital identities, but also created a framework to help people think about virtual identity. In a piece she wrote for *Wired* magazine, she argued:

> *The anonymity of MUDs gives people the chance to express multiple and often unexplored aspects of the self, to play with their identity and to try out new ones. MUDs make possible the creation of an identity so fluid and multiple that it strains the limits of the notion. Identity, after all, refers to the sameness between two qualities, in this case between a person and his or her persona. But in MUDs, one can be many.*

In some cases, such as experiencing life as an African American or Hispanic avatar, different identities can be therapeutic and beneficial. In others, such as adults posing as children, they can be dangerous.

Turkle's discussion of identity-switching in online textual worlds foreshadowed today's graphic, 3-D, online worlds in which people "try on" avatars as if they are shopping for a new pair of shoes. But how does using a transformed identity via an avatar change the person behind the avatar? If a person wears an avatar that is different from herself in some way—for example, younger, shorter, or thinner—does that person change her views of her self?

As the figure below illustrates, a man can easily see himself in the virtual mirror as a male avatar, a female avatar, or even as a chinchilla! The equipment tracks his actions (for example, his hand touching the area near the virtual mirror) and renders those actions on an avatar of any form imaginable.

Nick Yee took his first major step toward a career studying virtual reality while he was an undergraduate at Haverford College, by writing an honors thesis reporting his analysis of the identities people use when playing online games such as *EverQuest* and *World of WarCraft*—describing who they were, what they did, and how they represented themselves online. Throughout graduate school and after earning his Ph.D. at Stanford University, he continues to explore this fascinating topic. In the first Proteus Effect experi-

Who am I? A man can take on any avatar imaginable in virtual reality.

ment, Yee brought participants into the virtual-reality laboratory, where, donning a head-mounted display, they walked around a virtual room, encountering a virtual mirror (as illustrated in the figure below). In a typical mirror study, he asked participants to spend about ninety seconds gesturing in front of the mirror and observe their "reflection" as it moved with them. The mirror image, while sometimes looking very much like the participant, at other times was mismatched in a strategic way. For example, in Yee's first study, the avatar's height had been altered relative to the participant's actual physical height.

Based on numerous studies, psychologists know that height is positively correlated with leadership, confidence in social situations, and even income. For example, consider recent research on the brutal world of online dating. In a landmark study, Günter Hitsch, a professor in the business school at the University of Chicago, investigated the economics of online dating. He and his colleagues reviewed profiles of approximately 22,000 online daters for a three-month period in 2003. They examined features of the profiles (for example, height, weight, income levels, and body type) as well as the success of the dater (for example, the number of times other users accessed the dater's profile, examined photos of the dater, and sent e-mails to the dater). The researchers correlated the features of daters' profiles with how desirable the daters were. Using mathematical models, they determined how the features combined with one another to produce a desirable date. For men, height has tremendous implications in the world of online dating. According to the authors, "A man who is 5 feet 6 inches tall, for example, needs an additional $175,000 in income to be as desirable as a man who is approximately 6 feet tall (the median height in our sample) and makes $62,500 per year."

Yee sought to understand whether virtual height could influence people in the same way. To do so, he examined how individuals negotiate with one another in virtual reality. He found that participants with taller avatars out-negotiated those with shorter avatars. Regardless of their actual physical height, people with virtually shorter avatars were three times as likely (compared to people with normal-size or taller avatars) to accept an unfair settlement. It is important to note that in the height manipulation, the magnitude of the height difference was very small. After the study, participants were asked to describe the purpose of the experiment and to describe their avatars. Not a single one reported that their avatar was taller or shorter than the other avatars in the virtual world. Evidently, even though participants did not consciously realize that their avatars were shorter or taller in virtual reality, they exhibited psychological effects typically associated with their virtual stature. Furthermore, when Yee statistically accounted for participants' actual height in physical space, it did not change the results at all! Regardless of one's physical height, changes in avatar height changed their behavior. The critical finding is that participants' self-perceptions changed their behavior.

Surprisingly, the confidence instilled by taller avatars persisted outside of virtual reality. Yee determined this using a clever design in which two participants were given height-mismatched avatars in virtual reality, after which they negotiated face-to-face. During the latter negotiation, they sat in chairs adjusted vertically to ensure their height was equal in the physical world (regardless of their actual height). The results of the physical negotiation were effectively identical to the results of the virtual negotiation from the prior study. If a person's avatar height was increased virtually, he was a more successful negotiator afterward in the physical world.

A similar effect occurs when virtual attractiveness changes. In one experiment, participants spent twenty minutes as avatars in virtual reality talking to other avatars about their personal lives. Some participants' avatars were slightly more attractive than average, while others were slightly less attractive. Participants in the attractive-avatar condition exhibited more confident nonverbal behaviors than the unattractive ones. While wearing attractive avatars, they walked three feet closer and entered the personal space of the other person. Moreover, they spoke more confidently. We recorded the conversation. Later on, judges who had no idea if they were listening to the attractive or unattractive avatars recorded how many personal disclosures the participant offered. Participants with attractive avatars divulged more personal information about themselves than the others. Surprisingly, a small change to how one's avatar looks changes the way he chooses to speak—those who wore unattractive avatars further hampered themselves socially by acting aloof.

In a follow-up study, Yee demonstrated that participant confidence acquired in virtual reality actually carried over to the physical world. After participants had worn attractive or unattractive avatars in virtual reality for about twenty minutes, they took part in a second, "unrelated" study in the physical world, which began about twenty-five minutes after the conclusion of the virtual-reality phase. In the "unrelated study," participants completed an online dating profile and arranged blind dates.

When participants chose people on the dating site who they thought would be interested in going out with them, those who had had more attractive avatars in the prior study selected better-looking people ("10s") than participants who spent twenty minutes in virtual reality as less attractive avatars. In other words, having an attractive avatar in the virtual world boosted their self-perceived attractiveness. That boost persisted outside of the virtual world—such participants actually thought they had a shot with better-looking people.

Yee's data also allowed comparison to the findings of a recent study by Jeffrey Hancock and his colleagues at Cornell University. These researchers examined how much people lie about themselves when filling out online dating profiles. Hancock examined profiles on a popular online dating Web site, and then contacted the daters, paying them a lot of money to actually visit his lab. Upon their arrival, the researchers weighed them, checked their height, took pictures to record how attractive they were, and then compared those values to the information listed on their profiles. It turned out that just about everyone lied. Most relevant here is that men tended to misrepresent their height, claiming that they were, on average, a half-inch taller than their physical height. While the height discrepancies, on average, were small, for many in the sample the difference was larger, up to three or four inches.

In a follow-up study by Yee on avatar attractiveness, after the virtual-reality experience, participants recorded their height, weight, and other personal characteristics in a mock online-dating profile. Tracking data collected during the previous, virtual-reality phase of the study allowed the team to determine their actual height, and, just like Hancock, he calculated the reported-versus-measured height discrepancies. Yee's study replicated the Cornell findings with two exceptions. First, both men and women, not just the guys, overstated their height. Second, and more important, when participants saw themselves as attractive in the virtual mirror, they didn't lie about their height afterward. We believe this happened because of the confidence participants gained from having had an attractive avatar. In other words, their virtual personae carried over to the physical realm, and they did not feel the need to exaggerate.

CHAPTER SEVEN

RE-CREATING YOURSELF

If notions of the self are malleable and susceptible to influence, then people use virtual reality to their advantage. Let's start with an inescapably human concern—age.

At Stanford, the Center on Longevity employs a group of scientists whose research focuses on the quality of life for older people. The director of the center, Laura Carstensen, and her protégé, Hal Ersner-Hershfield, are worried that young people are not saving enough money for retirement. Most eighteen-year-olds cannot fathom what it would be like to be thirty, let alone sixty-five. There is a consensus among financial analysts that young people today will be in dire straits later on in life if they don't change their savings and investing habits.

In the spirit of Yee's findings on height and attractiveness, we worked with the scientists at the center to determine if showing college freshmen sixty-five-year-old versions of themselves would increase their connection to the future, inducing them to save more.

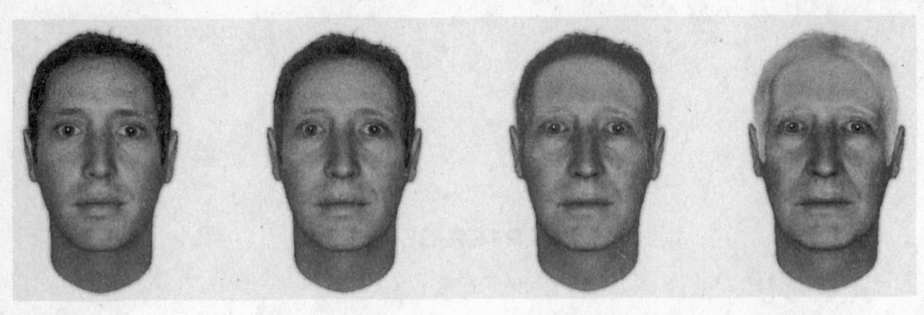

Young at heart: Hal Erner-Hershfield as he might look at different points in his life.

For a typical study, we photographed college-age participants, built 3-D models of their heads, and "aged" them graphically. The avatars above depict how Ersner-Hershfield, the lead scientist on this project, might appear at thirty-five, forty-five, fifty-five, and sixty-five. He's thirty-one now.

Participants entered virtual reality and viewed themselves in a virtual mirror either at their current or elderly age. They then had a twenty-minute conversation about their life in the virtual world, with another person. After they left the virtual world, we asked them how much money they would allocate toward retirement. Across three separate experiments, people in the elderly condition budgeted more than twice as much money for retirement as people in the young condition.

A DIFFERENT KIND OF "FAT SUIT"

Obesity is arguably one of the most significant problems facing the health of the population of the United States. In the United States, losing weight is one of the most sought-after goals, but also one of the most difficult to achieve. In virtual reality, however, weight loss is as easy as a keystroke.

William Gibson wrote in the opening chapter of *Neuromancer,* describing why Ratz the bartender didn't change his appearance, "In an age of affordable beauty, there was something heraldic about his lack of it." In what is one of the only longitudinal studies of avatars, we examined the implications of Ratz's decision. In a virtual world, where everyone is beautiful, what are the psychological implications of wearing an obese avatar?

To answer this question, we tested the effect of avatar body size on people's conception of their physical selves. Over six weeks, seventy-six students spent almost forty hours inside the virtual world of *Second Life*. Their task was to socialize with other avatars for at least six hours per week in the online world, with one catch—their avatars, unlike almost every one of the hundreds of thousands of avatars in *Second Life*, were built to be extremely obese. The figure below depicts the typical "default" avatar that *Second Life* provides, and the virtually obese versions of these avatars.

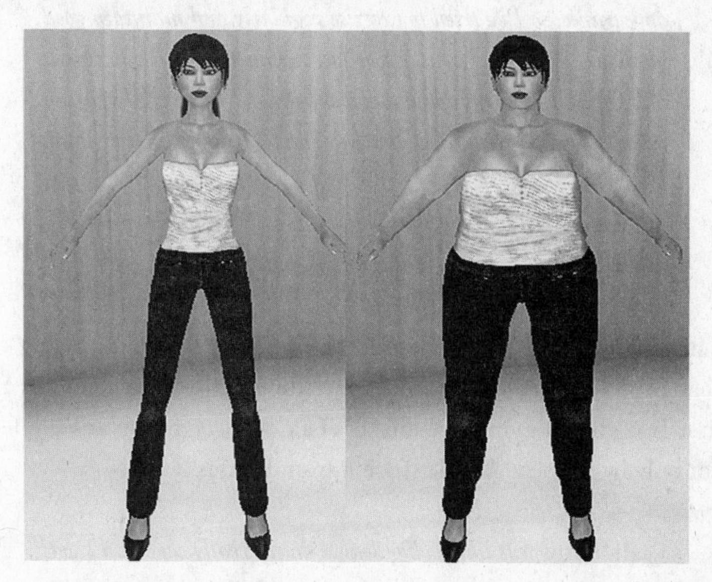

Supersize me! A typical Second Life *avatar and her obese counterpart.*

In a virtual world, where perfect bodies are not only free but the norm, we sought to determine the psychological consequences of being a physical outcast. Moreover, this study enabled us to examine the consequences of virtual obesity over time. The most telling data are the self-reports from some of our participants. To illustrate, after six weeks of interacting via an obese avatar, a college-age woman from New York writes:

> *When we were assigned a specific body shape at the start of our time in the world, I was fixed with the "overweight female" shape and was not allowed to modify anything in the body, torso, or legs sections of my avatar's appearance. It was ironic that I was given this type of shape, considering how conscious I am of my physical appearance. In the past, I have been heavier than the average person my age and have had to work extremely hard to lose weight. I was made fun of sometimes in high school because of my appearance, so I've tried to exercise regularly and maintain good physique. When my avatar was fat, I immediately felt conscious and felt as if I was fat again and was keen on changing this part of my appearance, though I knew I couldn't. At first, I didn't think it that big of a deal, because it wasn't actually me and it was just my avatar, and I didn't think she could really get made fun of in a world like* Second Life. *I was unpleasantly surprised.*

This student went on to describe how she spent hours trying to change her virtual body shape, including substantial amounts of time in a virtual gym. The effects of an overweight avatar were felt not only by women. A male student from Kansas wrote:

> *At first it was like no big deal. I didn't really care and I was just kind of hanging out and it seemed like the other people*

didn't care. I guess more importantly, but then like as I went to clubs and events and stuff, it was almost like they were like stereotyping me as a bigger guy, so, I don't know that's just how I took it.

According to this student, the negative consequences of an overweight avatar were exacerbated over time.

This idea of virtual identity can also be extended to race—though the concept of virtual race is a tricky one. Early on, scholars such as Sherry Turkle argued that on the Internet, stereotypes such as race will become obsolete, because digital identity transcends physical appearance. However, race is still a huge issue in virtual reality. A student in our *Second Life* experiment, described above, wrote:

> *One of the strangest things that I felt going through* Second Life *was that I was really one of the only black avatars pretty much everywhere I went . . . As I started to realize that I was literally one of the only black people on* Second Life, *I started to wonder what everybody else thought about the only token black guy walking around by himself. I didn't know the race or gender of the person across from me; seeing that they could have easily been black or Chinese and of a different race if they chose so, but I knew I was a black guy, and that made me second guess on some of the stuff I could do . . . because I thought all of these things, I don't approach people as much as I should, unless they start conversation first, it is likely that I just walk right by them without saying anything.*

In an experiment on avatar race, we worked closely with graduate student Victoria Groom. She thought that racial empathy would be induced when white participants wore black avatars. She reasoned that if white people "walk in the shoes" of a black person, their negative racial stereotypes would break down. She ran about one hundred participants, half of whom approached a virtual mirror in a black avatar and half in a white avatar.

Unexpectedly, she found the opposite—wearing a black avatar actually caused people to score higher on standard measures of racism than those who wore a white avatar. In other words, wearing a black avatar primed more racial stereotypes instead of creating empathy. Stunningly, this pattern was true not only for white participants but for black ones as well. Regarding virtual racism, it seems that the story is complicated. Although the research on face-to-face contact with members of out-groups demonstrates that taking a stigmatized other's perspective can reduce racism, this study indicates that an assigned racial identity actually makes the stereotypes more salient. Similarly, other scientists have developed simulations to study stereotypes, such as those for people with schizophrenia. Over a relatively long time in such a "mirror" world, the initial saliency of the stereotypes might actually lead to a lessening of them. But that hypothesis remains to be tested.

In sum, virtual worlds offer an unprecedented opportunity to separate people from their physical identity, and to role-play in a variety of manners. However, the role-playing is not "free," and actually has consequences not only with regard to online behavior but also for behaviors carried over into the physical world.

VIRTUAL DOPPELGÄNGERS

In *Being John Malkovich*, actor John Malkovich (playing himself) wakes up in a restaurant and looks across his table. He sees a seated woman wearing a revealing evening dress, but as his gaze pans up, he is stunned to see his own head on top of the voluptuous female form. Seconds later, a waiter walks by, also wearing Malkovich's head. John's emotional response is predictably dire, and his terror only increases as he pans the room and realizes that every single person in the restaurant—ranging from jazz singers to a midget—is wearing his head instead of their own. Malkovich is literally trapped in a roomful of clones (from the neck up) behaving independently of his own intentions and actions.

But with digital representations of humans, the notion of a doppelgänger suddenly becomes nonfiction. Sony's PlayStation video game *Tony Hawk's Underground 2*, or *THUG 2*, has sold more than one million copies in the United States alone. Players can upload their photographs onto the face of a character. Subsequently, via game controllers, players can make their virtual clones perform death-defying feats, such as skateboarding at dizzying heights.

Independently of game platforms, once an avatar is built, it can be animated in any manner fathomable! In Yee's studies on the Proteus Effect, avatars moved and talked like the person they represented, even though they looked different—for example, taller, more attractive, or older. But imagine a scenario where an avatar looks just like the wearer but *behaves* independently, by virtue of computer control—in other words, as an agent. In a series of studies, we investigated the consequences of people witnessing "themselves" performing actions in virtual reality. How does it affect someone's psyche to watch his avatar (i.e., "himself") behave autonomously?

Psychological theories suggest what will happen when people encounter their virtual doppelgänger. One of the most relevant is a classic "social learning theory," developed by Albert Bandura in the late 1970s, which posits that people model behaviors they see others perform. In Bandura's famous "Bobo Doll" study at Stanford in the late 1960s, children spontaneously beat up "Bobo" after watching adults do the same thing. Since then, researchers have demonstrated that the more similar the target is to the observer, the more likely the observer is to mimic that target. For example, people are more likely to imitate the behaviors of others who are the same sex, race, age, and even those who share their opinions. The virtual doppelgänger demonstrates the ultimate power of virtual reality—the target is not merely similar to the observer, it *is* the observer.

Can a virtual doppelgänger be used to encourage people to behave in a healthy manner? Jesse Fox, while a student in our lab at Stanford, ran a number of studies designed to answer that question. People know that healthy behaviors are important. They are barraged by the media on a daily basis, reminding them, for example, to exercise, eat right, and get a good night's sleep. However, relatively few people actually make the effort to do so. Why not? Some researchers posit that it results from a failure of "self-efficacy." People simply do not believe that they have the ability to lose weight or get in shape. However, by observing one's doppelgänger become the healthy person that one wants to be, it might be possible to encourage and induce healthy behaviors.

In one study, adults came to our lab and stood watching their virtual doppelgängers exercise or stand still. Participants in the exercise condition reported higher "self-efficacy"—the belief that they could exercise successfully—than participants in the still condition. Simply by watching their doppelgängers exercise, participants expressed the intention to adopt that behavior. More important, they claimed to have acted on their intentions: when we contacted all participants the next day, the exercise group reported that they had worked out more than participants in the still group.

In a follow-up study, participants physically exercised while wearing a head-mounted display. As they exercised, they watched their virtual doppelgänger losing weight. For every minute they ran in place in the physical world, their doppelgänger lost virtual weight. In other words, participants were able to connect the cause and effect of exercise and weight loss. Next, without compensation, participants had the opportunity to remain in the lab longer and work out using free weights. Participants who were in the critical, self-efficacy condition, exercised ten times longer than participants in a number of control conditions.

Jesse Fox reported similar effects on healthy eating behaviors. Undergraduates viewed doppelgängers eating candy and gaining weight. Subsequently, participants remained in the room to complete questionnaires with a bowl of candy nearby. The virtual eating behavior subsequently affected actual eating. However, the effects varied by gender—men ate more candy after seeing their doppelgängers (compared to control conditions), while women suppressed the urge and consumed less candy.

Doppelgängers can also affect memory. Kathryn Segovia, another graduate student at Stanford, ran one of the more industrious studies in the history of virtual-reality research, placing fragile, expensive equipment on the heads of scores of preschoolers and elementary-school children. Segovia was interested in false memories. There is a large body of work on false testimony and eyewitness memory. Much of it has focused on children, because they are more suggestible than adults. Unfortunately, innocent people have been imprisoned because of children's false memories. Previous work has demonstrated that if a child actively thinks about doing something, the action can often become real to the child when he's asked about it later.

Segovia's study examined whether witnessing a doppelgänger engage in a certain behavior could produce a false memory. She built replica avatars of sixty children, who then watched "themselves" swimming with whales in a virtual aquarium.

When interviewed after the study, more than half of the elementary-school children who saw their doppelgängers swimming, compared to other children who did not, persisted in the belief that they'd been to SeaWorld and had gone swimming with whales. Apparently, watching a doppelgänger can create false memories. The implications are profound. In virtual reality, just seeing oneself performing some action can change one's behavior and memory. Given

I endorse this product?

that virtual behavior is controlled by computer programmers and animators, the consequences can be dire.

One such consequence is depicted in Steven Spielberg's adaptation of the Philip K. Dick short story *Minority Report*. Specifically, there was a scene in which Tom Cruise's character looked up at a billboard and encountered an advertisement with his name in it. That marketing feat can certainly be re-created in virtual reality. We've demonstrated that if a participant sees his avatar wearing a certain brand of clothing, he is more likely to recall and prefer that brand. In other words, if one observes his avatar as a product endorser (the ultimate form of targeted advertising), he is more likely to embrace the product.

To further explore the consequences of viewing one's virtual doppelgänger, we ran a simple experiment using digitally manipulated photographs. We used imaging software to place participants' heads on people depicted in billboards using fictitious brands, such as the example shown on the previous page.

Sometime after the study, participants expressed a preference for the brand, even though they knew their faces had been placed in the advertisement. In other words, even though it was clearly a gimmick, using the digital self to promote a product is effective.

MOVING FORWARD WITH DOPPELGÄNGERS

When a virtual human is controlled by an actual human's behaviors in real time, it is an avatar, by definition. If it is controlled by a computer, it is an agent. However, the examples we describe in this chapter challenge these definitions. The Proteus Effect begs the question, "Is it still my avatar if it looks nothing like me?" The Virtual Doppelgänger begs the question, "If it looks just like me in every way but is controlled by a computer algorithm, is it really just an agent?" Strictly speaking, the answer to both of these questions is "yes." However, the notion of self is fluid in virtual reality. The same holds for grounded reality but to a lesser extent. "I wasn't myself when I said that!" is an excuse people often give for some transgression. The possibilities for self-representations are many in virtual reality, and scholars are just beginning to scratch the surface of understanding the implications of these new technologies for who we are and how we act as humans.

We have been exploring some of the practical implications of this work with the military. As a team leader and research psychologist at the Fort Knox Research Unit of the U.S. Army Research Institute, Carl Lickteig is innovating the way the army trains personnel to fight counterinsurgencies. U.S. soldiers in places like the Middle East have a difficult time understanding the culture and the manners of the people with whom they are interacting. For example, currently in Iraq, U.S. soldiers spend much time with Iraqi civilians, such as local governing council members. The problem is that, while it is possible to train soldiers via military manuals, there is no substitute to being there. Virtual reality, however, allows a U.S. soldier to "walk a mile" in the shoes of an Iraqi civilian, thereby giving the soldier a better understanding of his life. Collaborating with others from different cultures requires soldiers to understand and use correct cultural gestures, to remain calm in stressful interactions with council members, and to interact with local civilians in respectful ways.

We have designed simulations for the military using the virtual mirror described earlier. The general idea is that a soldier can take on the visual perspective of a local council member and understand the demands placed on him. In other words, the trainee enters a virtual simulation, such as a governing council meeting, and sees himself embodied as an Iraqi council leader. Just as in Nick Yee's Proteus studies, the soldier becomes someone else, this time an Iraqi council leader. Currently, many soldiers feel that local council members often make unrealistic demands during council meetings. One possibility is that virtual perspective-taking task may help the soldiers understand the pressures and demands placed on council members by their respective local constituents. Currently, U.S. soldiers do their best to put themselves in the shoes of the council members; however, according to Nick Yee's research, using imagination often is not sufficient for giving someone an alternative identity. Consequently, we are hopeful that virtual-reality simulations will increase the ability of U.S. soldiers to understand and be more empathic toward Iraqi civilians.

Even outside the world of games and avatars, people loosely identify with certain details of their digital representations. The Hancock study discussed earlier indicates that people routinely lie about their weight, height, and body type in online dating Web sites. Indeed, the ease with which people can self-present in various manners is what makes online communication unique. As the popular cartoon from the *New Yorker* points out, "On the Internet, nobody knows you're a dog."

Imagine a future in which thousands of experiences, day after day, occur virtually. Think about the first minutes after waking from a particularly resonant dream, or that first palpitation of awkwardness in an exchange with a person you dreamed about the night before. What happens when we don't have a "clean break" between the virtual and the real? What happens when we can reenter those worlds, and can do so willfully? Whether exercising and losing weight, swimming with whales, or watching ourselves endorse products in virtual reality, the remarkable capacity of this technology to alter our own perceptions and beliefs about ourselves makes what is thought of as "virtual" very "real" indeed.

CHAPTER EIGHT

STREET SMARTS

A DAY IN THE LIFE . . .

Dave had a meeting at eight A.M. He felt completely alert the instant upon waking, and it took him only seconds to get into his brand-new suit. Los Angeles traffic wasn't a concern, and he arrived at the conference room perfectly on time, where he sat with the rest of the corporate team.

He'd decided the best path to getting a raise was the age-old strategy of flattering his two bosses, Mrs. Robinson and Mr. Figgs. Robinson was demonstrative, warm, and liked her subordinates to participate actively in meetings. Figgs was old-school—stoic and authoritative, and treated his subordinates like children who were to be seen and not heard. At the meeting, Dave played the chameleon, making eye contact with Mrs. Robinson and smiling widely as he affirmed all her ideas, while showing deference in front of Mr. Figgs with pensive head-tilts and silent note-taking. Later, Robinson and Figgs, who were sitting side by side during the meeting, would agree that Dave was an up-and-coming prospect for greater responsibility.

For about a half hour during the three-hour meeting, Dave was able to take a nap. After work, Dave went out for happy hour on a first date with Jana. The date couldn't have gone any better at first—he was perfectly "in tune" with her conversational flow, his clothes were the perfect hue to match her eyes, and he even had a facial structure similar to hers, which he'd heard would subconsciously help cement the bond between them. Unfortunately, Dave caught a bit of a bug, not sure where he picked it up, and fell to pieces at the end of the date. Realizing there was no second date in the near future, he left the restaurant and hightailed it over to see his therapist, who was always on call.

. . . OR IS IT?

Let's rewind to what really happened that day. Physically, Dave slept past eight A.M., in fact, past nine A.M. Luckily, his avatar, like all avatars, didn't need sleep and joined the rest of the corporate team in virtual reality.

Unlike Dave, the rest of the team members were in control of their avatars in real time. Dave, lazy as he was, had his avatar on autopilot. Dressed impeccably in a digital Italian suit, the avatar was programmed to be a perfect participant. It laughed at jokes (taking cues from voice inflection changes of the other avatars), nodded in all the right places, and dutifully recorded the details of the discussion.

Mrs. Robinson and Mr. Figgs each saw a different version of Dave's virtual counterpart. After all, Dave's avatar, which sent itself over the network separately to all of the participants in the meeting, was programmed to appear different to each of them—amiable and open to Mrs. Robinson, quiet and respectful to Mr. Figgs.

At ten A.M., Dave lumbered out of bed and jacked into his virtual-reality system to check on his avatar, which, truth be told, made a far better impression in the business meeting than Dave could have ever done himself. He quickly scanned the meeting notes recorded by his avatar and realized that his avatar could probably advance his career better than he could.

That taken care of, Dave left his avatar on autopilot, making it, for all intents and purposes, an agent. While napping, he dreamed about the virtual date he'd made for the evening. Dave had spent a major chunk of his last paycheck on a "behavioral matching" program. Basically, during his virtual date, a computer algorithm would assess his date—her personality type, her emotional state, even the shape of her body—and would automatically adjust Dave's avatar's appearance and behavior to fit her ideal version of a man. Of course, it had to be subtle, but that's why he paid top dollar for the program. He did not want to commit a first-date gaffe.

But the world, and technology in particular, is far from perfect. Somehow the new software had some quirky network security settings, and a computer virus leaked through. Two-thirds of the way through the date, Dave's avatar became transparent, literally becoming the mirror image of Jana, as opposed to subconsciously matching her. Jana, suddenly finding herself across the table from another woman, was not pleased, and it took her only a minute to realize Dave was using digital augmentation software (which she'd specifically listed as a "pet peeve" in her dating profile). "Another phony," she thought. Disgruntled, dejected, and suddenly alone in the virtual bar, Dave settled up and ported over to a therapist for a ten P.M. emergency session about the trials and tribulations of virtual dating.

∞ ∞ ∞

Being gregarious creatures, humans seek the company of others. We spend hours each day monitoring—often unconsciously—the way others react to us; meanwhile, we are constantly being examined, in turn. In the waning moments of a first date, for example, one likely would take careful inventory of the person across the dinner table. Subtle behaviors such as eye contact, voice intonation, and posture provide clues about their level of attraction—and the chance they will say "yes" to a second date. Other times, these assessments are automatic, such as mindlessly keeping appropriate distance from strangers on a subway platform.

Successful people in social situations tend to be sensitive and appropriately responsive to verbal and nonverbal cues. Social responsiveness has been studied for centuries by philosophers and psychologists, but in the past fifteen years it has received greater scientific attention. Notably, Yale social psychologist Peter Salovey identifies "emotional intelligence" as the key to social responsiveness. Similarly, Tufts University's Robert Sternberg has argued that "street smarts"—the ability to adequately assess and display social responsiveness—is a cornerstone of intelligence, one that equips us to operate successfully in the world.

Let's look a bit more closely at some of these crucial but unwritten rules of social engagement. On the nonverbal level, maintaining eye contact with another portrays confidence. But staring at the eyes of another person for too long (try doing this for more than seven seconds to a stranger) will overwhelm the recipient. Verbal conversation functions similarly: people who talk too much are annoying, but so is the silent treatment. Over a lifetime, people develop and learn the implicit social rules of responsiveness to a greater or lesser degree.

These rules are highly complex and contextually bound. For

example, a warm, enthusiastic embrace of another may make sense at a family reunion but not so much at a summit of heads of state. For many people, the rules of responsiveness are difficult to master—some people can effortlessly "work a room" while others struggle to find an appropriate way to introduce themselves at social gatherings. For some, such reticence is diagnostic of clinical disorders, such as autism. Even those who have a highly developed level of emotional intelligence are not perfect, and even the most astute social players, such as some politicians, are not completely responsive all the time.

Emotional intelligence can reach new plateaus in virtual reality. Avatars and agents can be made much more socially responsive via artificial-intelligence algorithms. The ability of programmers to create perfectly socially responsive virtual humans is only limited by the knowledge of the social rules alluded to above. While building the perfect artificially intelligent virtual human may be decades off, the work we discuss in this chapter shows that even small steps in responsiveness can lead to huge gains in social influence. Programmers can augment agents and avatars with a "street smarts" algorithm, programmed to recognize every verbal and nonverbal nuance from everyone in the virtual world, and to respond appropriately. On a date in the physical world, a woman might not be able to laugh at every corny joke made by her companion. Her avatar, however, can fill in and provide the appropriate laughter at his lackluster musings. The possibilities are endless.

HERE'S LOOKING AT YOU

As teachers, we give a lot of thought to creating the best possible student-teacher interactions. Consider eye contact again. Other than physical touch, looking another in the eye is the most potent non-verbal tool humans possess. The psychology literature is filled with dozens of eye-contact experiments. We know that if you look another person in the eye, his heart beats faster, he pays attention to what you say, will remember the interaction, and will likely buy what you are selling.

When a teacher looks at a student, the student has a better chance of actually learning. This makes sense in a one-on-one setting, but loses effectiveness in a lecture hall with a hundred students. A teacher can try to spread her gaze around, but even if she spreads it evenly among the students (which is almost impossible to do while speaking), she can only gaze at each student on average 1 percent of the time.

However, when teaching a class in virtual reality, the teacher can employ augmented gaze powers. Because the rendering computer in a virtual-reality system, such as those we have described above, sends information to all of the students' systems individually, the information can be tailored to each student. So, instead of sending the position of the teacher's eyes and head veridically, the system can be programmed to simultaneously display to each student the illusion that she is looking directly at him or her. Moreover, each student will believe that she is the sole recipient of that gaze. In other words, digital avatars can enjoy the ability to engage in mutual gaze with everyone at the same time, a feat that can't be replicated in the physical world.

We have run hundreds of experimental participants in what we call "non-zero sum gaze" virtual classroom studies. Thus far, not a single student has ever detected that the gaze of their teachers' avatars was not genuine. Moreover, their classroom behavior improved dramatically. As a group, they paid more attention to the teacher's avatar than those whose virtual teachers did not include augmented gaze. They also retained more information. In sum, using relatively simple code (about twenty lines), one can render the behavior of a virtual teacher in socially responsive ways. This augmentation can positively change students' behaviors, attitudes, and learning rates.

As university professors, we know well that teacher-student interactions are not the only ones that take place in a classroom. There are student-to-student interactions as well, sometimes in whispers but more often nonverbally. If a group of students in a classroom starts looking away from the teacher, many others will mimic that behavior to see for themselves what may be happening. In virtual reality, this can be turned into a didactic advantage.

Imagine two extreme types of student gaze—one in which all the students are hanging on every word of the teacher and looking directly at her, and one in which all the students are ignoring the teacher, looking away from her and staring off into space. In which class would you rather have your children? We examined the possibility of manipulating not only the gaze of the teacher but the gaze behavior of one's fellow students—in this case, a seminar conducted in virtual reality. In our experiment, student participants were placed in a virtual conference room with ten embodied agents of other "students" sitting around an oval seminar table. We placed the virtual teacher agent at the head of the table. The figure on the following page reveals the point of view of a participant sitting at the seminar table, looking at the teacher, who is standing. Note that in this image, the other students are looking at the face of the teacher while she conducts the seminar. Half the participants in the study had a group of attentive classmates—that is, fellow students who looked at the teacher during 80 percent of the lecture. The other half had distracted class members who looked at the teacher only 20 percent of the time.

The results of this study demonstrated that it was possible to in-

An attentive audience in the virtual classroom.

crease students' learning merely by transforming the attentiveness of their fellow classmates to the teacher. Participants in the "high-gaze" condition looked at the teacher more (i.e., paid attention), and their performance on memory tests later showed greater retention than that of students in the "low-gaze" condition. Virtual reality thus allowed us to optimize a learning environment, and we've learned that even small changes to the gaze of virtual others in the room can profoundly influence behavior.

It is also possible to use virtual-reality technology to alert teachers to their own classroom behavior. It is extremely difficult for a teacher to spread his or her eye-gaze around equally to all of the members of a class—it's hard enough to look at only two other people at an equal rate while talking, let alone an entire classroom full of people. We ran a study using a simple algorithm—when the teacher ignored a student for too long, the student literally started to fade before the teacher's eyes, as the figure on the following page illustrates.

In this virtual-reality study, the more translucent a student's avatar appeared, the longer he or she had been ignored by the teacher.

Disappearing without a glance: A student fades before the teacher's eyes.

Without the aid of the visualization algorithm, students on the far edges of the room got ignored—defined as not being looked at by the teacher for a fifteen-second interval—for approximately 40 percent of the lecture. However, when teachers were able to sense the disappearing students, the percentage decreased to about 10. Once again, a simple transformation using virtual-reality technology allowed the teacher to improve his performance and increase learning by class members as a group.

In a follow-up study, we collaborated with Peter Mundy, a renowned scientist who studies autism at the University of California Davis. Mundy hypothesized that our visualization algorithm would help autistic children, who typically have a difficult time looking others in the eyes. However, if others in virtual reality faded when the children did not look at them, the autistic children behaved "normally"; that is, looked the others in the eyes in a manner similar to that of non-autistic children. Based on these findings, we believe there is a great potential for using this technology to help autistic children learn how to navigate the social world.

KNOWING ME, KNOWING YOU

In a related vein, let's consider imitation. Many psychological studies have demonstrated the power of mimicry. Typically, people unconsciously mimic the nonverbal behaviors of an individual with whom they are interacting, displaying similar postures and gestures, which generates a more favorable impression of the mimickers by the party they are mimicking. Politicians know this, and often try to mimic their constituents—recall Hilary Clinton's attempt at a Southern drawl when she spoke to a church group in Alabama during the 2008 primaries.

Recall that Duke professor Tanya Chartrand coined the term *chameleon effect* to describe the social benefits of mimicry. One of Chartrand's colleagues, Rick van Baaren, who has been working with her for over a decade, runs the "Unconscious Lab" at Radboud University in Holland. In a conceptual replication of Chartrand's work in a physical-world setting, Van Baaren examined tipping behavior in restaurants. He instructed servers to repeat the order verbally as they wrote it down or to merely write it down and acknowledge the order. Servers that repeated the order verbally received tips that were almost twice as large as the control conditions! This is particularly notable given that, in the Netherlands, tipping is not as customary as it is in the United States.

Now consider that avatars have tools human mimickers can only dream about. Recall how a virtual conversation functions—John in New York smiles, and his computer sends the digital tracking data of his smile to Ellen in Wichita. In near real time, her computer receives the information and renders his avatar smiling. Upon receiving this information, Ellen winks in response, and her computer detects this, sending this information to John's computer that then renders her wink via her avatar. In essence, this is the same way that a normal telephone interaction works. But virtual reality provides social interactants with super-mimicry powers. Ellen, or even a third party, can simply set her avatar's digital controller in such a way that it mimics John's nonverbal facial expressions and gestures automatically. In other words, John sees his own smile played back on Ellen's avatar. Ellen doesn't need to actively make each gesture—the virtual system receives all of the tracking data and puts that data on her own avatar automatically, without Ellen having to hit any buttons or even think about doing it.

In several studies, we have demonstrated that if Ellen's avatar is superpowered in this way, John will like her more. In one, participants listened to a speech by an agent in virtual reality. The agent took an unpopular position in the speech, at least in the minds of the participant pool. The agent attempted to persuade students that they should sign a petition about a security policy on campus that required all students to always have their student identification on their persons. In one condition, the agents mimicked the head movements of the participants by repeating them about four seconds after they occurred. We had previously determined that at a four-second lag, very few people consciously detected that their movements were being mimicked. In the other condition, participants saw the agents' head movements but they weren't repeats of their own. In fact, they saw the movements from a participant in one of the previous sessions, who had been mimicked. Consequently, across the two conditions—mimicry and no mimicry—the movements were exactly the same, but for half of the participants, the movements were their own, rendered on the virtual agent (their conversational partner) four seconds after they occurred.

Three notable findings stood out. First, less than 5 percent of the participants had any idea that they were being mimicked. We succeeded in making the manipulation "implicit." Second, in the mimicry condition, the participants maintained eye contact with the agent for longer durations than in the control condition. Finally, the participants were more likely to sign off on the unpopular policy when the speaker mimicked them than when the speaker did not. In sum, mimicry was not detected consciously but influenced the listener's nonverbal behavior and decision-making in predictable directions.

In a second study, we extended this work to virtual touch. People put a lot of stock in handshakes—dozens of books have been written on the topic, and many classes in business schools actually teach techniques on how to master a confident, firm grip. Often, the first step in a social interaction is the handshake, and as Kevin Eikenberry, business consultant and author of a number of business etiquette books, points out, "A good handshake is firm but not overpowering. It isn't the precursor to a wrestling match, and it doesn't feel like a dead fish . . . Always make your grip firm, but make adjustments based on the firmness of the other person's grip." In our study, we put Eikenberry's advice to the test: instead of adjusting to the firmness of another person's handshake, we created avatars who exactly mimicked the handshake of another person. We used a "force-feedback haptic device" that basically records the movement and force one uses when moving the device with his or her hand and plays it back mechanically.

In this study, two experimental participants arrived at the lab, where we recorded each of their handshakes using the force-feedback device. They then shook hands virtually, by feeling the mechanical recording of each other's handshake. Except there was a catch: one of the participants did not get the other person's handshake. Instead, he received a recording of his own handshake, believing it was the handshake of the other person. In other words, one of the two people shook his own hand! Our prediction was that the participant who unwittingly mimicked the handshake of the other would gain social influence.

To test this, we used a standard negotiation structure, in which the two participants debated rent rates and other parameters; one took the role of a landlord while the other took the role of the renter. Our results demonstrated that handshake mimicry was effective: participants who mimicked the others were liked more than participants who did not. Once again, this occurred despite the fact that not a single one of the participants consciously detected that their handshake was being mimicked. More important, we demonstrated negotiation effects. Participants who were mimicked reported liking their partners more and were less aggressive in the negotiation. The use of mimicry actually affected how much money they gained in the negotiation. Interestingly, this effect was largely driven by male participants, which is not surprising, as social norms dictate that the handshake is a stereotypically masculine behavior. Females, on the other hand, were not as affected by the nonverbal mimicry.

In these two studies, it was rare for anyone to consciously detect the mimicry, but we wanted to understand the consequences of "getting caught" using virtual mimicry. In some ways, mimicry is an offensive behavior. For example, kids tend to taunt one another by engaging in deliberate mimicry, repeating everything that another says in an effort to annoy them. In this sense, mimicry can become mockery. To this end, we designed an additional study that made the mimicry more obvious, by programming the target head-movements to be more deliberate and noticeable. We demonstrated that when mimicry is detected, social influence backfires completely. Specifically, virtual agents that were "caught" mimicking participants received less support for the security-identification policy, compared to the control condition, while agents who were not detected received increased support for the policy. In this sense, mimicry is effective, but when the receiver becomes informed of it, she can defend against this persuasion strategy.

The best part of virtual mimicry is that it frees one up to focus on other tasks. So, on a virtual date, Ellen can maximize her charms by using automatic, scripted nonverbal mimicry, and she can then dedicate more of her cognitive resources to devising clever and witty things to say. Likewise, in a large virtual classroom, a teacher can automatically mimic all her students simultaneously with no effort, whereas in the physical world, she can only mimic one student at a time and is forced to devote substantial cognitive resources to the mimicry. Virtual mimicry, in other words, accrues more benefits than physical mimicry can, while taking much less mental effort to implement.

BENDING DISTANCE

In 1905, Albert Einstein wrote about the theoretical possibility of warping time and space over large distances. In some ways, that vision is fulfilled by virtual worlds. By its very nature, virtual reality creates the illusion of closeness: avatars can feel as if they are in the same room, while their owners are physically positioned on separate continents. But in the same way that people's responsiveness can be increased by selectively transforming their avatars, in virtual reality, the laws of physics can be avoided or even reversed to allow each avatar to tailor space to their advantage.

Consider the following assumption. In a large lecture room, students that sit in a "sweet spot"—for example, in the center of the room a few rows in front of the podium—are thought to learn better than students who sit in other places. In a series of studies, we randomly assigned students to various seats in a virtual classroom and confirmed that this was indeed the case; students sitting in the front/center position scored about ten percentage points higher than other students when tested on the material they learned while sitting in the lecture room. In a typical university class of a hundred students in a physical lecture hall, only a few students can sit in the preferred location. But in virtual reality, it is possible to render the geographical positioning of the room separately for each student. Consequently, in a class of a hundred students, each student can perceive that she is sitting in the preferred seat and is alone in that seat. Person A sees Person B in the back of the room, and Person B sees Person A in the back of the room, while both of them perceive themselves front and center! The amount of learning gain is substantial with only two people, as we have demonstrated in a number of experiments in the laboratory. But imagine the gains one could accrue in a larger class (for example, at the University of Michigan we took a course that had more than a thousand students). These simple "geographical transformations" have the potential to drastically change distance-learning.

Indeed, virtual space can be bent in ways that humans have never experienced physically—consider the notion of placing two bodies in overlapping virtual space. In a series of experiments, we collaborated with University of California, Berkeley, professor of Electrical Engineering Ruzena Bajcsy. Professor Bajcsy is one of the world's foremost scientists in computer vision and artificial intelligence, and she runs one of the country's most advanced virtual-reality labs, where she has designed a system to track and render a person's full-body movements in real time. Instead of tracking only a small number of movements on the body, she re-creates every single joint and movement. Our studies examined whether one could learn physical motions, in particular the martial art tai chi, better from a virtual avatar with whom one could share body space than from a traditional video tutorial. In two separate experiments, learners came into the lab and were instructed on how to perform a sequence of tai chi moves they had never tried before. Half of them saw a two-dimensional representation of a recorded teacher (similar to a video tape), while the others saw a virtual room that contained a real-time rendering of their own avatar as well as a 3-D recording of a teacher. They could literally insert their virtual bodies inside that of the virtual teacher agent. In other words, participants could learn the moves by keeping their own avatar's limbs within the confines of the teacher's limbs during the performance. Later on, the students performed the tai chi moves from memory in front of a camera. Once recorded, we sent the videos of the performance to a tai chi expert to judge the moves. Students who learned virtually performed substantially better than students who did not have the ability to break the rules of physics and enter the teacher's body space.

Some may object that virtual augmentation of social responsiveness is fundamentally disingenuous, that it unleashes possibilities for new kinds of deceit. Indeed it does (and we'll tackle that issue in a later chapter). But it's our view that humans largely pursue new technologies—such as virtual reality—in good faith, and that we're inclined to create and to employ such technologies for beneficial ends. Think of how many times we wonder if "someone took it the wrong way." Ask yourself if you would enable a mimicry or responsiveness algorithm if you were the student in a class or a participant in a crucial virtual corporate-strategy meeting. By enabling such algorithms, you would absorb and retain more information. Assume you knew that the avatar next to you, who wants that A or that promotion as much as you do, was doing it. Would you?

CHAPTER NINE

ETERNAL LIFE

IN HIS PULITZER PRIZE–WINNING BOOK *DENIAL OF DEATH*, PHIlosopher Ernest Becker argued that humans are terrified of their own mortality and, consequently, avoid the topic. As perhaps the only species "gifted" with conscious awareness of impending death, escaping the inevitable has been a goal of humans throughout history. But, with few exceptions, humans never really escape "death anxiety." Sooner or later, in grounded reality, death stares us in the face.

We argued in the early chapters of this book that humans have developed tools—ranging from storytelling to virtual reality—that facilitate mind-wandering and, in turn, evade existential anxiety. Indeed, death and immortality are major recurrent story themes across all media.

For example, the *Epic of Gilgamesh*, one of the oldest ancient poems, was first recorded in ancient Mesopotamia in the seventh century BCE, but scholars believe it arose from legends dating two millennia earlier. This epic poem describes its hero's intention to live forever after suffering through a friend's death. And indeed, three thousand years later, much of contemporary art and entertainment examines the same issues, from Tony Soprano's existential neurosis to the immortal vampires of the *Twilight* series.

Most religions posit immortality via some sort of "afterlife." Philosophers from ancient to modern times have grappled with and debated humans' relationship to their own mortality. Scientists are no exception. Not surprisingly, much research focuses on extending human life. In the last century or so, advances in personal hygiene, diet, exercise, promoting healthy habits, avoiding stress, medicines, vaccines, and surgery have extended the average life span remarkably for much of the human race. In the U.S. economy alone, the amount of money that is dedicated to stalling and masking the aging process is staggering. Billions of dollars are being spent on research to slow biological processes associated with aging. A prime example involves research on the shortening of DNA strands ("telomeres") that occurs whenever a cell replicates. Such shortening is thought to be a key to reversing the biological aging process.

Hundreds of billions of dollars have been spent masking the visual, auditory, and somatic effects of growing old. Most has been spent on such things as makeup, plastic surgery, lotions and potions, tacitly or not so tacitly applying the principles of self-perception theory—making people "look like a million to feel like a million." But significant sums have also been spent on wearable devices, such as glasses, contacts, and hearing aids. Literally going deeper to lengthen life and its quality, surgeries such as Lasik, age-spot removal, and even human-to-human organ transplants have become routine. Finally, synthetic body replacement parts—such as hip and knee joints, kidney dialysis machines, artificial hearts, and even synthetic blood—range from routine to experimental at this point. However, none of these investments guarantees anything approaching immortality.

Scientific opinion about the feasibility of immortality itself ranges from "impossible" to "achievable in the next few decades." The latter potential is the quest of scientists in disciplines ranging from medicine to robotics to philosophy. The promise of their work has created a market for businesses that offer to preserve people's entire bodies, heads, brains, or even stem cells cryogenically or otherwise, so that they can be restored to life if and when such a promise is met. Perhaps thousands of people around the world have taken advantage of such services, though the reliability of such offerings is questionable. So far, no one has restored a cryogenically preserved human body to life. Additionally, one wonders about the condition of current cryogenically preserved bodies and body parts if bringing them back to life becomes technically possible in the future. For example, the *New York Daily News* has reported that baseball legend Ted Williams's cryogenically frozen head, regrettably, has been battered with a monkey wrench.

If and when it becomes possible, one way or another, to re-create a living body via cloning, or otherwise replicating the genome as an individual person, major problems would remain. For one thing, genes interact with the environment to a great extent. The inevitably different environment of the future will likely moderate the individuals' personalities, abilities, motivations, etc. For example, memories for certain things, such as goal-oriented body movements (think golf swing) and expressions (think smiles) result from the interplay of skeletal muscle and brain. Athletes talk about such muscle memory all the time. Would Ted Williams really be able to play baseball again if his body were somehow re-created? With a new body, would he even know who he was?

Because of these and other problems related to the preservation of the human body, a number of scholars have begun discussing the necessity of replicating consciousness. For example, in the future, a supercomputer might be able to exhaustively encode the various neural connections in a person's brain before it's frozen or cloned. Imagining the technology is daunting, as the brain is thought to have upward of 200 billion neurons with each connecting (synapsing) on average to 10,000 other neurons. That's 2 quadrillion (2,000,000,000,000,000!) connections in all. It'd be a formidable task, made more complicated by the fact that these connections are said to be "plastic," meaning they are constantly disconnecting and reconnecting.

All of these techniques are ethically controversial. Fortunately or unfortunately, however, we have no proof at all that any of them will actually work. Nevertheless, one of the more optimistic voices in this discussion is Ray Kurzweil, the noted author and futurist, who believes that these types of systems should be available by about 2040.

In his book *The Singularity Is Near: When Humans Transcend Biology*, he notes:

> *We have the means right now to live long enough to live forever. Existing knowledge can be aggressively applied to dramatically slow down aging processes so we can still be in vital health when the more radical life-extending therapies from biotechnology and nanotechnology become available. But most baby boomers won't make it because they are unaware of the accelerating aging process in their bodies and the opportunity to intervene.*

We will probably have to leave verification of Kurzweil's prediction to those who follow us in life.

Why is this discussion relevant to this book? In our humble opinion, there is another path people may be interested in taking, one we call *virtual immortality*. The notion of virtual immortality differs from the notion of preserving consciousness. The idea is that, with virtual "tracking data" collected over a long period of time, one can preserve much or even most of people's idiosyncrasies, including a large set of behaviors, attitudes, actions, appearances, etc. One will not be able to "relive" life through an avatar, but nonetheless, a digital being that looks, talks, gestures, and behaves as they once did can occupy virtual space indefinitely. In this sense, there are two ways to think about "immortality." One is extending the nature of one's life to be able to continue to enjoy the fruits of living. The other, less experiential, is about preserving one's legacy. Having a version of oneself around forever allows a person to affect future events and shape the experiences of others. Historically, media have allowed certain privileged people in the past to do so (e.g., Solomon, Jesus, Mohammed, da Vinci, Michelangelo, Jefferson, etc., etc., etc.). Virtual reality will enable nearly everyone to do so.

Regarding preservation of one's legacy, consider "terror management theory." The theory was first conceived in the late 1980s by University of Kansas graduate students Sheldon Solomon, Jeff Greenberg, and Tom Pyszczynski. This team of researchers and others has provided much evidence that death anxiety, whether conscious or unconscious, is what motivates many kinds of human behaviors. We, too, believe that such existential anxiety is at the heart of humans' penchant for psychological travel away from grounded reality to virtual reality.

In a typical experiment, participants are "primed" with thoughts of death—sometimes unconsciously, sometimes more consciously—that make their own mortality more salient. In some studies, words such as *death* or *funeral* or pictures of their own face are flashed to people without their awareness of these stimuli. In other experiments, primes are environmental, without attention being drawn to them, such as driving people around in cars and slowly passing by funeral homes, instilling an implicit awareness of an unavoidable death. Once these subconscious fears are primed, participants behave in predictable ways to cope with death anxiety.

One of the primary defense mechanisms to combat these subconscious reminders of mortality is to excessively support the culture to which one belongs. In laboratory studies, subconsciously terrified American participants display huge pro-American biases. They even punish people described as violating cultural norms, or go way out of their way to help people they know to be "patriots." The idea is that supporting their cultural worldview is a way to make them immortal by proxy. In other words, they would attempt to bolster America while they're alive; since they're a part of America, so long as the stars and stripes survive, then a part of them lives on as well. The terrified subconscious provides itself with symbolic immortality by identifying with their country or culture, entities that are both larger and longer-lasting than the transient life of a single human. The theorists argue that fear of mortality drives many of our day-to-day social behaviors, from how we treat people we meet to the laws we pass and the clothes we wear. Some have even suggested that inducing fear in citizens, such as increasing the number and severity of terror alerts, helps incumbents get reelected.

Moviegoers might recall when Marty McFly traveled in time in *Back to the Future* and got caught up in his own parents' romance. The movie plays upon a simple desire most of us have—to know our parents when they were young and childless. In virtual reality, people may soon be able to "travel back in time" to realize that desire. People might also recall the time Marty ended up in the Old West (two movies later) and ran across his great-great-grandfather. Identifying with Marty, again, many people would welcome the opportunity to talk to long-since-passed ancestors. But perhaps even more appealing is to look at the interaction from the perspective of Marty's great-great-grandfather. Who wouldn't love to know that long after we're dead, we'd still be able to touch the lives of our progeny generations down the road? It is, in short, a ticket to a virtual immortality.

As described above, in the physical world, many people do their best to extend longevity with multivitamins, proper diet, and regular trips to the doctor's office. However, if immortality is the goal, then we're currently out of luck biologically. On the other hand, ten minutes inside a typical virtual-reality setup allows digital-tracking equipment to capture literally millions of bytes of data about a person's movements, appearance, and behavior. In less than two hours, if a system records behaviors at high resolution, the tracking data will overflow hard drives on most computers. Consider today's typical young adult, who spends more than twenty hours per week online, year after year. The amount of data that can be archived about a single person is astronomical.

Moreover, virtual-reality technicians can build a socially interactive version of anyone. A 3-D digital model provides a near-perfect analog of a person's body and face (indeed, we have used them for studying police lineups and eyewitness testimony, as we discuss later). Motion-capture technology allows the acquisition of one's gestures. Sooner or later, artificial-intelligence technology will permit implementation of one's personality traits and other psychological idiosyncrasies. Storage of every possible phoneme in one's language, and in one's own voice, will enable one's eternal avatar to say things the physical self never even said. So-called haptic devices capture one's touch—how firm a handshake is, the way one hugs, and how one moves her hands. The overall result will be an avatar that looks and behaves like the person it represents, but can do so even when that person is no longer alive, sort of like Disney World's Abraham

Lincoln but with artificial intelligence and social-interaction capabilities. After one passes on, his great-great-great-grandchildren can enter a "holodeck," sit on the long-deceased ancestor's lap, tell him about their day, experience his avatar tell a story, give a hug, and provide advice. A quite reasonable facsimile of a person's dynamic tendencies can be preserved indefinitely in virtual reality.

ALIVE FOREVER

In 2001, we began a research project at the University of California, Santa Barbara, and called it "Alive Forever." We put an ad in the local paper and recruited participants. Our idea was to build each person's avatar and to allow them to keep digital versions of themselves stored forever, perhaps to interact with their distant progeny. Our plan, as psychologists, was to study the reactions and the ways people invested themselves in their digital counterparts. Given the research findings from terror-management theory, demonstrating that people become more reliant on in-groups and culture when their mortality is threatened, we were curious to find out whether it was possible to "reverse" that trend by convincing people they could live forever in digital space. Consequently, we constructed digital versions of about fifty middle-aged experimental participants recruited from the community and convinced them that their avatars would be preserved in a University of California–sponsored digital vault for generations to come. We were surprised by some unanticipated findings.

In all psychology experiments, it is required and important to "debrief" the participants, that is, to let them know the reasons behind the study and take them through all the details of the experiment, including its purpose and hypotheses. For the first time since creating a virtual-reality lab, we experienced outrage from some of our participants. Some actually became so irate after learning that it was just a psychology experiment (and that we weren't actually going to store their digital selves "for centuries") that we had to end the study only a few months after it began. Even after we told the participants that we couldn't store the avatars, or that we couldn't even build them yet to be exact replicas with the technology we had in the lab, a number of them proceeded to contact us long after the study was over in order to find out if we still had their avatars. Obviously, people took the notion of storing their digital selves very seriously.

But a virtual avatar is not just personality, and appearance does matter. Consider the case of Orville Redenbacher, the noted popcorn mogul, one of the most successful brand icons in U.S. history, and frequent proclaimer that "You'll taste the difference or my name isn't Orville Redenbacher" during his many appearances in television commercials advertising his product. Mr. Redenbacher passed away in 1995, but there was enough video and photographic material from decades of his public life to build a 3-D model of his head and body, and to use sound clips to create a persuasive analog of his voice. Consequently, in a surprising move, in 2006, the owner of the brand, ConAgra, released a commercial with a virtual Orville Redenbacher dancing around with headphones on, popping popcorn, and exclaiming, "Can you believe this little baby holds thirty gigs?" as he held up an iPod.

The avatar-like spokesman was attempting to make a comparison to the lightness of his popcorn. Other commercials in the past spliced video together by taking actors out of context—for example, by overlaying a Dirt Devil vacuum cleaner on Fred Astaire's dance partner. However, Redenbacher was perhaps the first major example of someone reanimating a completely novel action on a reconstructed 3-D model of a deceased public figure. It may be the case that Mr. Redenbacher would have been excited about the idea; perhaps he even endorsed similar ideas before passing away. His grandson, Gary, claimed that "Grandpa would go for it. He was a cutting-edge guy. This is a way to honor his legacy."

On the other hand, Orville Redenbacher may have been completely mortified by the word choice and dance moves fathomed by the commercial's director. Not everyone is excited about being digitized. For example, Electronic Arts produces a college football game that features characters who bear an uncanny likeness to the students on the actual teams. Even though the players are not named in the game, some of them feel that the physical likeness is close enough to bring a lawsuit against Electronic Arts. As Dixie Flatline, a character whose personality was revived in virtual space in William Gibson's famous novel *Neuromancer,* said, "I wanna be erased."

The ethical controversy surrounding the possible animation of one's avatar in nefarious ways has received ample attention in the legal community. For example, three decades after his death, John Lennon appeared in a manipulated public service announcement for One Laptop per Child, proclaiming:

> *Imagine every child, no matter where in the world they were, could access a universe of knowledge. They would have a chance to learn, to dream, to achieve anything they want. I tried to do it through my music, but now you can do it in a very different way. You can give a child a laptop, and more than imagine, you can change the world.*

In this instance, his widow, Yoko Ono, gave permission to the advertising agency to carry out the pro bono work, but, of course, there was massive outrage expressed by fans over the manipulation of the dead artist's image and voice.

Consider the influx of virtual actors, or, as some call them, "synthespians," we see today in films. Perhaps the most interesting recent depiction of virtual actors was the film *S1m0ne*, starring Al Pacino and directed by Andrew Niccol. Pacino plays a director who gets sick of managing flaky, undependable actors, and resorts to virtual versions of specific actors. The avatars always show up on time and obviously do whatever they are programmed to do. Virtual reality completely removes the "diva factor" from acting.

The notion of digital actors is not just fiction. Consider the impressive list of actors who already have detailed 3-D scans of their heads and even bodies: for example, Brad Pitt, Arnold Schwarzenegger, Jim Carrey, Michelle Pfeiffer, Denzel Washington, and Gillian Anderson. In the making of *Polar Express*, Tom Hanks's movements were meticulously tracked and recorded in order to construct appropriate graphics. Would the director, Robert Zemeckis, be able to use those data in future film projects, no longer needing the actor himself?

A clever headline in the *Los Angeles Times* proclaimed, "Old Actors Never Die; They Just Get Digitized." The technology to resurrect three-dimensional, animatable models of deceased actors from film archives is available and beginning to be utilized. For example, a startup in Los Angeles, Virtual Celebrity Productions, has already acquired the exclusive worldwide rights to represent in digital form Sammy Davis Jr., James Cagney, Marlene Dietrich, Vincent Price, George Burns, W. C. Fields, and Groucho Marx.

Not only advertisers and actors are pursuing digital-identity capture. The National Science Foundation recently awarded a large grant to a team of scientists from the University of Illinois at Chicago and the University of Central Florida. These scholars were tasked with developing the technology to build a lifelike visual appearance and personality replica of specific individuals. The goal of the project is to develop the technology needed to accomplish such a task. To test the feasibility of the technology, they have chosen to capture, preserve, and reuse the expertise of retiring NSF program director, Dr. Alex Schwarzkopf. His doppelgänger is accurate in terms of appearance (they used a 3-D scanner to make the avatar look like its owner), nonverbal behavior (meticulously recorded via motion-capture technology), personality (by using a vast array of artificial-intelligence algorithms based on data collected during interviews with him), and emotions. One of the graduate students, who spent more than thirty hours per week with Schwarzkopf for a year, collecting data, noted that the doppelgänger even used speech idiosyncrasies, for example, his signature salutation, "Keep the peace!" In the figure below, Dr. Schwarzkopf's avatar mimics the expression of disgust from the actual Schwarzkopf.

This system is functional and is being used by current, less senior NSF officials to leverage the expertise that Schwarzkopf achieved over decades of service. He has given them carte blanche to use his avatar as they see fit. The avatar actually forms new relationships with the people it meets, and its understanding of the world at large changes based on its conversations. In this sense, virtual Schwarzkopf forms with physical people lasting relationships that physical Schwarzkopf never had.

We would be remiss to talk about virtual immortality without

Schwarzkopf and his doppelgänger.

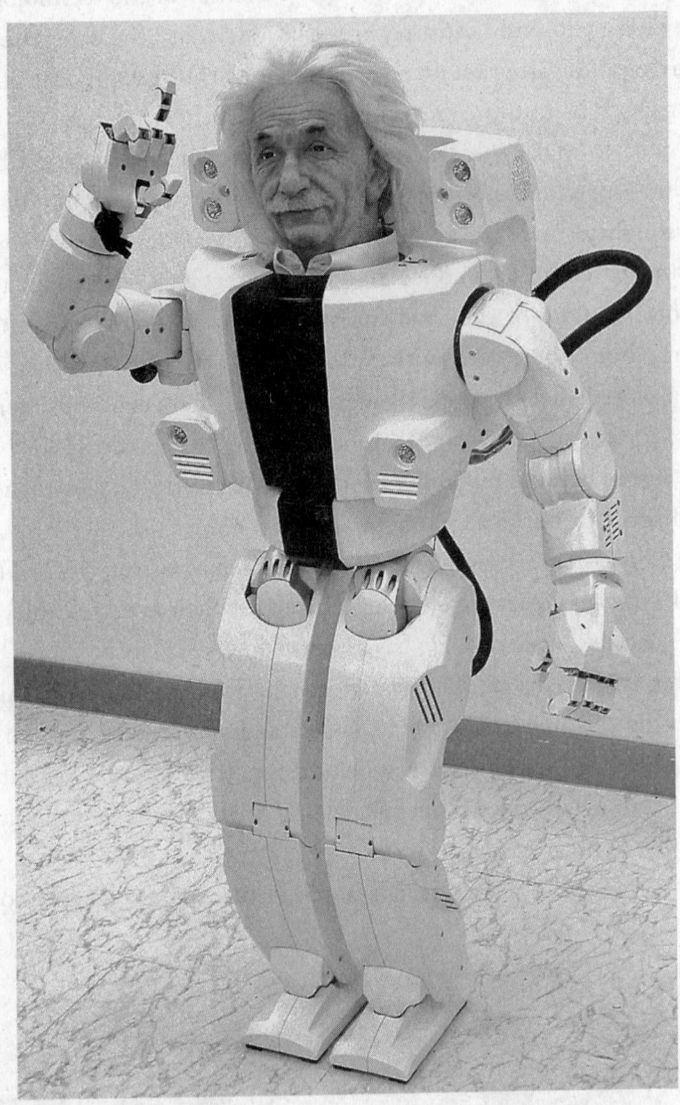

David Hanson's Einstein robot: a look at another possible manifestation of "virtual immortality." What if the robot had your face, voice, and personality?

discussing the David Hanson's robots, photorealistic versions of famous figures, including Albert Einstein, depicted opposite.

While at NextFest, a science exhibition sponsored by *Wired* in 2004, we had a booth set up next to Hanson. He displayed an animatronic robot that could gesture in response to people interacting with it; the gestures, flesh movements, and eye contact were quite realistic. The robot could respond in novel manners based on the situation, by using cameras to detect human action and then animate itself appropriately in response. The level of presence of the robot was stunning. It is rumored that one of the first robots built by Hanson, named Eva, was modeled after a girl he was dating. Obviously, the creation of Eva raises some fascinating and dangerous issues. What happens if someone builds a virtual version of you? Can you ever actually break up with him? Sure, you can physically leave, but there will always be that virtual version of you that can cater to any whim of your ex. Throughout history, lovers have saved mementos of one another; now it is possible to save *versions* of one another.

INFINITE REPLICABILITY

Think about how often people multitask nowadays. For better or for worse, we are often mentally in more than one place at the same time, or at least traveling back and forth between many places psychologically very quickly, whether text messaging in movie theaters or chatting on the phone while driving a car. Think about identical twins, who have the unique ability to send a proxy to a boring meeting or to take a difficult exam. Avatars and agents take the notion of multitasking to a new plateau, and may free us up a great deal, via a virtual power we label *outsourcing*. For example, movie hero Bruce Willis, the physical version at least, has an active social life, attending parties in Hollywood and New York with other celebrities. But while Mr. Willis was enjoying his lunches and glitzy affairs in the summer of 2007, his agent clones were promoting the movie *Die Hard 4*. A doppelgänger of Bruce Willis was built and replicated. An army of these doppelgängers were sent into *Second Life* to walk around and convince people to see the movie.

People are familiar with audio agents. For example, in 2006, an audio agent of Samuel Jackson randomly called people's cell phones to convince them that *Snakes on a Plane* was a film that would change their lives. In 2007, a virtual audio agent of Scarlett Johansson called thousands of California primary voters to convince them to vote for President Obama.

Not only can our avatars work while we sleep, but *thousands* of versions of our avatars can work simultaneously as we sleep (maybe, like Bruce Willis's, they're off performing a canned sales pitch to any number of potential clients). Imagine how much more persuasive, and useful, full-blown avatars and agents in virtual reality will be. The current body of research demonstrates that people who multitask perform worse than those who focus all of their attention on a single task. For example, drivers who talk on the phone tend to be dangerous; consequently, laws have been passed against using handheld devices while driving.

A recent discovery by Clifford Nass (of Microsoft's Clippy fame) and his colleagues examined high and low multitaskers. Those who regularly engage in a variety of media simultaneously—for example, texting, watching TV, Web surfing—did worse on focusing attention and storing information, and got distracted more often than infrequent multitaskers. However, in virtual reality, if one deploys a number of versions of her avatars, there is literally no cognitive cost, because it is up to the computer to monitor the avatars performing secondary tasks. Indeed, many online games today feature this concept of "secondary avatars," which are akin to smart pets or indentured robots, characters that are somewhat controlled.

The notion of having many versions of a personal avatar is raising warning flags both in government and industry. Recently, we were contacted by a high-level manager at a large defense contractor, who was pursuing projects to develop an "Avatar DNA" identification test, that is, a way to be sure that if one is talking to an avatar of a given person, he or she could be sure that the person matched the avatar. It is natural to assume that people will "borrow" the avatars of others, sometimes openly admitting to the mismatched occupation, other times using deception to leverage the identity of others. The defense contractor sought to develop algorithms that could ensure that an avatar actually was being controlled by its rightful owner, either by analyzing the movements and speech of the avatar or by some elegant, password-type system, similar to what is currently used by online banking.

All of these ideas reflect the fact that humans have been seeking immortality since the beginning of recorded history. For the first time, one can preserve large portions of identity via avatars. It may not be immortality exactly, but it is arguably the closest we can come, for now, to living forever. Some may find these ideas comforting, others may be disturbed. Either way, the evolution of agents and avatars over time will be fascinating to watch.

CHAPTER TEN

DIGITAL FOOTPRINTS

EVERYONE LEAVES FOOTPRINTS.

Robinson Crusoe used them to track down Friday, who, like *Cast Away*'s Wilson, could have been a figment of Crusoe's imagination. Forensic scientists use footprints to track down criminals. Examining length and width, they can estimate a person's height; by examining depth, they can estimate weight. Looking at the impression itself, for example the sole of a shoe, they can learn much about the person who left them. Women wear different shoes than men, construction workers wear different shoes than investment bankers. New shoes leave perfect patterns, worn shoes leave "scuffy" patterns. From that information, forensic scientists can draw conclusions about how wealthy someone is (i.e., are they wearing new or old shoes?). Like fingerprinting, they can match a shoe and its owner to the footprint. Even when the naked eye cannot see a footprint, ultraviolet lighting and other tools can reveal enough information to help identify a suspect.

Now, think digital "footprints." In the previous chapter, we discussed how every action and utterance of a person can be archived in order to preserve his legacy. How? Analogously to the footprints people leave in grounded reality, people leave traces in virtual reality. Every "step" one takes in a virtual world leaves "footprints" that are as detailed and predictive as any forensic scientist could hope. Recall that virtual-reality technology can track and render nearly all of a user's movements—the ones people "read" to draw conclusions about others' emotions, personality, age, etc. Such tracking creates archives of individual movements and social interactions that were never before possible.

Currently, conversations can be video-recorded for research or other purposes. But researchers don't often do so. Generally, people feel uncomfortable when they know a camera is present. In virtual reality, there are no visible cameras. However, "hidden cameras" can record everything in high definition from any viewpoint or distance, enabling the reproduction of every possible point of view without disrupting the virtual world or people in it.

Digitally recorded information is not limited to virtual reality. Many people are hesitant to admit how often they Google themselves. Some people have a huge Web presence, including images, videos, and personal profiles scattered online. Others have a smaller presence. But one thing is certain: Google never forgets. Google backs up, or "caches," the entire Internet frequently. Once something is online, it is practically impossible to erase all records of it. People may have some control over access to their digital information; for example, who can see their photographs on Facebook. However, they cannot control what other people do with this information. A person can limit access to her photographs to only a few people. But it takes only one person to e-mail it to someone else or post it on Picasa—once the cat is out of the bag, there is no putting it back.

Certainly, governmental security agencies are interested in digital footprints. However, Orwellian concerns arise when it comes to using them to identify who people are and what they plan to do—directly conflicting with their right to privacy. We return to this discussion of ethics later on, in chapter 13.

Nevertheless, the question remains regarding how confidently one can determine the identity of someone via their digital footprints. In the winter of 2008, we tried to find out ourselves, via a class project at Stanford. During the course, Stanford students learned about virtual worlds by spending time in one—the main course requirement was to spend at least six hours a week in the online world *Second Life*, and to write about their experiences. We anticipated about thirty people enrolling, but the idea of taking a course in a popular online virtual world appealed to the student body. More than three hundred students tried to register, but we were forced to limit enrollment to about eighty.

Our study was unique. Unlike other work examining digital footprints, we knew our participant pool well, in terms of their physical identity. We knew their age, their body-mass index, their race, their grade point average, their major, and their gender. We gave them a barrage of personality tests. So, we could track physical and virtual identities simultaneously in a rigorous, quantitative fashion during a longitudinal study.

Only the first class meeting was held in physical reality. In that session, students learned how to log in to *Second Life*, as very few had actually spent time inside the online virtual world. From then on, all class meetings were held in *Second Life*. Below is a screenshot of "office hours," in which teaching assistants demonstrated how to use the *Second Life* interface. The students are waiting online to receive their virtual tracking device, which we called *the sender*.

The sender was an invisible "attachment" to an avatar that re-

Office hours: Students wait to receive their "senders."

corded every virtual action participants made on a second-by-second basis: where they walked, which way their eyes were pointed, what they said, and all conceivable actions they could perform within *Second Life*. In other words, we had a record of every action eighty people performed in *Second Life* for six hours per week, for most of an academic quarter (six weeks). This dataset was enormous. By the end of the academic quarter, we accumulated massive amounts of tracking data from each person's sender every second. After weeks of data collection, we had complete, voluminous records of every participant's actions, including where they visited, whom they spent time with, and what they said.

We classified the students' demographic data into binary categories; for example, male/female, old/young, Caucasian/non-Caucasian, engineer/non-engineer, extrovert/introvert, high body-mass index/low body-mass index. Using mathematical formulas embedded in machine learning algorithms, we searched for subtle patterns in the data. The idea was to use the tracking data from the second-to-second virtual behavior to predict which demographic categories a given person fell into. We believed that these records about how people navigated in *Second Life*, the way they typed during chat, and whom they spent time with would give us clues to their personality and identities.

The results were striking. We were able to predict a participant's demographic and personality characteristics quite accurately (70 to 90 percent). To illustrate, the digital footprint data predicted participants' actual race, weight, height, math skills, exercise habits, attitudes toward women (e.g., sexism), and certain personality characteristics.

Think about the way clinical psychologists and psychiatrists assess personality disorders. During hours of therapy, a client's profile emerges from therapist-client interchanges. In a few months of weekly therapy, a typical therapist accumulates a notebook of data. The computer, on the other hand, accrues libraries' worth of data—not a single action goes unrecorded.

But it is not just the amount of data that makes virtual reality unique. It's the objectivity. A detective forms an impression about a suspect. In grounded reality, she brings her own biases and preconceived notions to bear and looks for verification of her notions. On the other hand, the analysis of virtual-world data can be completely unbiased. It records every behavior and looks for patterns via brute force. What makes such analysis superior is that machine learning algorithms can start with an unbiased "blank slate" each time.

Similar to detectives, clinical psychologists also must avoid biasing their evaluations. Albert "Skip" Rizzo, a psychologist and researcher at the Institute for Creative Technologies at the University of Southern California, developed a way to diagnose schoolchildren with attention deficit hyperactivity disorder (ADHD) using archived data generated and stored in virtual reality. To do so, he created an elementary-school classroom in virtual reality, where young children, donning head-mounted displays, sit at a desk in a classroom in which a teacher provides typical instruction to students. During the lesson, several distracting events occur, such as a classmate sailing a paper airplane across the room, a loud car passing by on the street outside the classroom window, and someone making a noise.

We've watched the playback of the head movements of children's avatars in the virtual classroom. It's a snap to identify a child with ADHD. Their tracked movements give them away, making the diagnosis obvious in three or four minutes. Kids with ADHD exhibit head and gaze movements that frequently meander. Kids without ADHD keep focused, for the most part, on the teacher. In our opinion, this simple paradigm could easily replace costly and time-consuming batteries of paper-and-pencil or even computer-administered tests currently necessary to make this diagnosis. Rizzo's pioneering work indicates that virtual reality will provide diagnostic methods that are accurate and cost-effective.

These studies are preliminary and modest. The samples are not large enough to draw universal conclusions. Given that we only examined Stanford students, and Rizzo only a few children, they are not as useful as large-scale studies involving people from many cultures. However, one thing is overwhelmingly clear from the experiment: digital footprints reveal much about physical and psychological identity. Actions that seem trivial in a virtual world—whether one chooses to walk or run, how quickly one types, or how close one stands to other people—provide clues about the self.

∞ ∞ ∞

DMITRI WILLIAMS AT THE UNIVERSITY OF SOUTHERN CALIFORNIA is one of the few people in the world who applies social science methods to study virtual worlds and online video games. His work with colleagues at a number of other universities utilizes one of the most comprehensive datasets of digital footprints to date. Williams has the "server data"—that is, the data from the company that makes the video game *EverQuest 2*, an online fantasy in which thousands of players from all over the world interact simultaneously via avatars. The server stores every action from every player in the game. These researchers also have gathered in-depth questionnaire data from about seven thousand players, and can link those data to the digital footprints of the players revealed via tracking.

Williams and his colleagues build computational models to predict players' physical attributes—for example, gender, age, and nationality—by pairing tracking data with players' demographics from the questionnaires. In one study, they were able to predict how likely one is to be a "high achiever," based on the tracked communication patterns.

Williams's team is one of the few that have access to "server data," and they are perhaps the only scientists that currently combine large server datasets with matching large questionnaire datasets. Game companies are extremely hesitant to release server data, because they don't want to violate the trust and privacy of players. Indeed, Williams had to sign a "double nondisclosure agreement." First, under no circumstances could the scientists learn the personal identity of the players, only their online account identification numbers. Second, the game company could not receive any findings concerning particular account numbers. In other words, the game company couldn't use specific player patterns to target advertisements or to kick them out of the game. In addition to privacy concerns, the magnitude of the data is a deterrent for most social scientists. In order to process all of the data, USC had to build a new cloud-computing center; that is, a massive storage and retrieval system.

WE'VE DISCUSSED DIGITAL FOOTPRINTS FROM VIRTUAL REALITY, online games, and social networking sites. However, it is also possible to predict identity and behavior from digital video. There is a saying that "the eyes are the window to the soul." For psychologists that window is the face. Since Darwin, scholars have analyzed the ways in which the face portrays emotion, though the past few decades have seen this work accelerate. Paul Ekman, arguably the father of experimental research on facial expressions and emotion, developed the well-known Facial Action Coding System (FACS), which has been used to study emotions, predict deception based on facial expressions, and has even inspired the dramatic television series *Lie to Me*, starring Tim Roth.

Ekman's system requires laborious manual coding of facial expressions—only trained and certified experts who have poured over videos frame-by-frame to learn FACS are qualified to make assessments. Moreover, scientists using an Ekman-type strategy are looking for predefined patterns of facial expressions; for example, smiles and frowns.

On the other hand, computer vision–based tracking systems can automatically quantify even the smallest nuances of facial movements. These systems code mammoth datasets at very little cost. We have worked for many years with Omron, a Japanese corporation that specializes in building sensing devices. Omron has developed OKAO Vision, the facial tracking system illustrated in the figure on the next page.

The white dots on the face in the figure are the landmark points

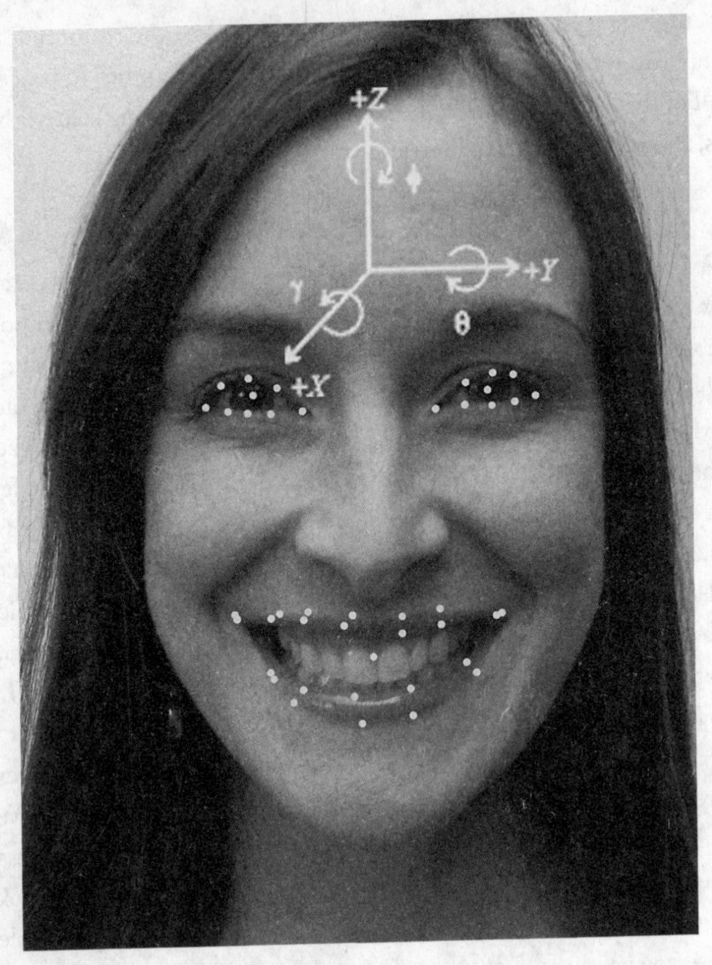

A facial expression tracking map.

that OKAO "tracks" via a Webcam. The observed individual doesn't need to wear anything on her face; the system uses algorithms involving shading and contour to detect the head direction as well as eye and mouth positions.

Facial expressions often predict ensuing behaviors—for example, blink rate, a proxy for fatigue, is correlated with poor driving behavior. Consequently, car companies such as Nissan and Toyota have supported our work using OKAO. The goal was to use a camera that could be mounted on a car dashboard (indeed, in Japan many cars now come equipped with this type of setup) to detect facial expressions in order to predict potential car crashes. In a series of experiments, we put people in a driving simulator, and tracked their facial movements while they drove. The simulator recorded car movements and data on a second-by-second basis; for example, speed, lateral movements (e.g., swerving out of a lane), compliance with ordinary traffic laws (e.g., ignoring a stop sign), collisions and accidents, etc.

The simulator automatically synchronized the car data with the facial tracking data, and using the mathematical models described in the *Second Life* study, we were able to distinguish the facial movements of good from bad drivers. For any given driver, the model accurately identified what we call the *pre-accident face*; that is, a specific sequence of facial movements that consistently preceded traffic accidents. Such models can separate good drivers from bad, and can predict an accident up to ten seconds before it occurs for any individual driver (with enough time for the car to take action and override the driver).

ANOTHER AREA WE EXAMINED IS HUMAN ERROR. IN FACTORIES across the world, vast numbers of injuries occur because attention wavers during tedious manual tasks. We designed a virtual assembly-line purposely built to cause boredom and fatigue. Experimental participants were required to pick a certain type of screw out of a box containing many types and insert them into designated holes in a virtual panel, with a haptic (i.e., virtual touch) device.

We video-recorded the workers' faces while they completed forty-five minutes of work. Inevitably, they made mistakes. We recorded each mistake and linked it to the facial tracking data. The results were strong. Facial movements predicted errors with high accuracy about five seconds before they occurred. Moreover, by looking at a few minutes of data from any given person, we predicted whether they were going to be a high- or low-quality worker, based on the overall number of errors they committed. The face not only told us when an error would occur, but also the type of person that was error-prone.

Of course, we are not the only people to use digital tracking data. One of the most respected groups researching this area is at M.I.T.'s Media Lab. Alex "Sandy" Pentland is the pioneer of what he labels *reality mining*. He has collected data using a plethora of digital devices ranging from cell phones to identification badges to online digital footprints.

For example, he observed ten male-female couples, ranging in age from twenty to forty, as they shopped together for furniture. The couples wore "sociometers," which recorded their speech intonations and grammar patterns as well as their movements and gestures. The devices also measured how the men and women interacted with each other; for example, how close they stood together and how synchronized their gestures were. Pentland and his colleagues then monitored whether the couples purchased furniture, and separated high-interest couples (e.g., ones who bought furniture) from low-interest ones, and looked for patterns in the tracking data to differentiate the two. By examining how "in sync" the couples were, the scientists could predict interest. Couples that were more balanced in speaking patterns were more likely to take a serious interest in products.

Jared Curhan, a professor in the business school at M.I.T., joined

Pentland in his work on reality mining. Curhan wondered if a thin slice of digital tracking data from a conversation could predict negotiation skill. In his study, fifty-six pairs of participants came to the lab to negotiate. In each dyad, one participant took the role of middle manager and the other the role of vice president. A computer program monitored the voices of the negotiators, paying attention to factors such as turn-taking structure and how synchronized the voices were in terms of intonation and prosody. As predicted, digitally tracked conversational dynamics occurring within the first five minutes of a negotiation were predictive of the outcome.

Pentland's group has succeeded brilliantly in proving that tracked data can provide clues about future behavior, but they are handcuffed by a major limitation: they need to design measurement devices and adorn their participants with them. Participants have to wear a sociometer. In virtual reality, however, every action can be tracked without such devices, establishing the gold standard for reality mining.

Clearly, the idea of using tracking data to predict behavior is not new. Psychologists have been mining language databases in recent years. For decades, James Pennebaker, a professor of psychology at the University of Texas, has been analyzing speaking patterns and word choices, using these data to predict one's identity. In one study, he instructed participants to either lie or tell the truth about whether they were "pro-life" or "pro-choice," information he already knew from a previous questionnaire. He found that by analyzing grammatical structure, he can detect deception. Truth-tellers used more first-person-singular words (e.g., *I*, *me*, *my*) than liars, and fewer negative-emotion words (e.g., *hate*, *worthless*, *enemy*). Pennebaker's group also uses grammatical structure to predict gender. They analyzed more than ten thousand text files from a number of previous studies, and demonstrated that women used more words related to psychological and social processes (e.g., *happy*, *grief*, *share*), while men referred more to object properties (e.g., *car*, *tree*, *red*) and impersonal topics. This linguistic method can predict a lot about a person—whether they are depressed, experiencing trauma, or even considering suicide.

In 2003, when the Iraq War began, the coalition forces used a deck of playing cards to identify all of Saddam Hussein's top henchmen. Imagine that you are a U.S. general in Iraq who has intercepted an e-mail or voice message sent from one enemy physical location to another. You only have the time and resources to attack one of the two places, and you want to focus on the one that has the highest command level, the "ace of spades," not the "jack of diamonds." Can linguistic analysis help? The National Science Foundation awarded Jeffrey Hancock and his colleagues a large grant to use digital footprints to determine how "high-ranking" a message sender is.

Hancock's team examined hundreds of memos that coalition forces captured from Saddam Hussein's administration. Each memo had a heading and a body. The heading indicated the status level (e.g., "private" v. "lieutenant"), and the body had language the scientists analyzed to find patterns that differentiated the two status categories. The models they developed successfully differentiated the two categories.

High-status commanders were less likely to use first-person pronouns (e.g., *I* and *me*) and more likely to write a loosely organized, rambling message. In contrast, lower-ranking individuals crafted well-formulated messages to their superiors. These findings are striking, given they were achieved by applying English-language models to Arabic messages. If the intelligence had been in English, the accuracy undoubtedly would be higher.

Hancock has used similar techniques to determine the digital footprint of "psychopaths." Using transcripts from interviews with Canadian murderers on death row, he segmented the prisoners into "normal killers" and "psychopaths" based on the clinical outcome of the in-depth interviews by psychiatrists. His automated linguistic analysis techniques demonstrated that psychotics were more likely to use words that denoted short-term hedonistic goals, such as food, drink, and money, while non-psychotics were more likely to reference higher-level concepts, such as family and religion. He posits that law-enforcement agents, in real-time pursuit, would be able to determine via a cell phone or e-mail exchange whether a suspect was psychotic, which would drastically change the strategies of engagement. Also, the digital footprints could be used to predict rates of recidivism.

And finally, on a much lighter note, online daters will be happy to know there is a tool that can detect those who lie about their age, weight, and height. Hancock and his team paid online daters to come to his lab. After archiving their profiles, he weighed them, measured their height, and examined their drivers' licenses. He then compared the actual data to the data from the online profile, and categorized the daters into "liars" and "truth-tellers." When looking at the text from the "About Me" descriptions in their profiles, he could differentiate the two categories via the linguistic structure of their profile. It turns out that liars show more "psychological distance": for example, using fewer first-person pronouns. In essence, using this model as a plug-in to the Web, one can run a program during online dating searches to determine how up front a potential date is. This takes speed-dating to a whole different level.

IN SUM, VIRTUAL REALITY PROVIDES "FOOTPRINTS" THAT CAN REveal who a person is, what they plan on doing, and whether they will succeed. While users in online worlds believe they are totally anonymous, nothing could be further from the truth. A virtual footprint tells us much about identity. There is a modern adage about Internet users, summed up in a 1993 *New Yorker* cartoon by Peter Steiner: "On the Internet, nobody knows you're a dog." Quite the contrary, say what one will, but with the bounty of tracking data, there is no way to hide the fact that one is a dog. Unlike with physical footprints left in the sand, there are no waves to wash away digital ones.

CHAPTER ELEVEN

THE VIRTUAL "JONES"

Throw moderation to the winds, and the greatest pleasures bring the greatest pains.

—DEMOCRITUS

Too much of a good thing is not enough.

—MAE WEST

ALTHOUGH DEMOCRITUS RECOGNIZED THE PITFALLS OF EXCESSIVENESS almost 2,500 years ago, he could never have imagined the endless possibilities for virtual excess that exist today. Contemporary temptations, in particular, make Mae West's joke a sort of "spiraling law" of behavior, in which yesterday's excess becomes today's moderation and today's excess becomes tomorrow's moderation and so on and so on, creating new compulsions and an ever-rising threshold for hedonistic satiation. In the 1960s and 1970s, parents worried about their children watching too much television and going to rock concerts. As parents ourselves, we worry about our children being mesmerized by online activities such as games, social networks, etc. But intergenerational concern runs both ways. Children have cause to worry about their parents' online activities. More and more frequent news reports describe children neglected by parents who are themselves obsessed with online activities.

Today, digital media provide new tools that allow people to act in ways, good and bad, that have never been as numerous or as accessible in the past. Ponder the following example. Yair Amichai-Hamburger received his Ph.D. in Intergroup Conflict from Oxford University, and has been studying the societal impact of the Internet for more than a decade. He and his colleagues at the Interdisciplinary Center in Israel are dedicated to exploring the positive and negative aspects of time spent online.

One project they describe is run by a Canadian mother and her son for a Zambian service organization whose mission is improving the lives of homeless children. Volunteering from home, mom and son use the Internet to identify, contact, and arrange logistics for potential donors anywhere in the world seeking to help the African children. True, volunteering isn't anything new, and the Internet is just one more tool to make volunteering easier and more effective. However, this case is especially noteworthy because the son has cerebral palsy with spastic quadriplegia and is confined to a wheelchair. But he still can access the Internet and give of his time, illustrating that people with special needs can volunteer via the Internet in ways they would be unable to do offline. In virtual reality, the son could even play soccer with those children.

In some ways, technology has complicated life at work. Think about multitasking, a seemingly critical skill for today's managers. A whole industry has evolved to design and produce digital instruments to facilitate multitasking. These "tools"—laptops, wireless Internet, iPods, iPads, Kindles, BlackBerries, iPhones, Bluetooth headsets, touch screens, etc.—have changed leisure time for many into work time.

These devices ostensibly "improve the standard of living." Indeed, the screen size of family-room televisions has steadily grown larger, and the picture definition higher, over the years. The variety of broadcasts has also increased exponentially. One can simultaneously watch television or even Web broadcasts of concurrent Major League Baseball, National Football League, National Hockey League, or National Basketball Association games. Owning the newest, biggest, most gadgety television screen shouts affluence to others.

History reveals that technological achievements create new human "needs" that sooner or later lead to increased consumer demand. Such demand leads to the development of even "better" technologies, creating even greater perceived needs. This cycle can create a seemingly vital but sometimes vicious circle, especially for technophiles. How many people felt the need to have a cell phone twenty or even ten years ago? How many people needed a Kindle or an iPad just a few years ago? How many people feel such needs now? Digital technologies typically do not lose their capabilities so much as become obsolete because of newer, more powerful ones.

Indeed, Jaron Lanier, who remains a huge proponent of virtual reality as an experience, argues in his recent book *You Are Not a Gadget* that pervasive piecemeal advances in digital technology are taking humanity down the wrong path. He maintains that instead of dedicating efforts to expressing true creativity resulting in wonderful content, developers are obsessed with quick response times. As Lanier points out, "You have to be someone before you can share yourself."

People are surrounded by increasingly compelling virtual technologies: high-definition displays such as television monitors, surround-sound systems, virtual camera angles, multimedia communication devices combining voice, video, text, etc., and digital 3-D movie projection. Wii game system tracking devices make digital media more responsive to players' body movements, thereby creating a much more immersive experience. Microsoft's Xbox 360 with its Kinect tracking uses infrared cameras to "infer" what a user is doing. Kinect allows people to play games naturally without any sort of game controller by tracking players' gestures, spoken commands, and even the objects they hold, such as a pen, paper, or baseball bat. The battle continues for domination of the game-console market as more and more vendors become convinced of the value of increasing the immersive nature of player experiences.

What's the problem with all this?

There is a steep and deep downside to ever-increasing digital-media capabilities. Many people have a penchant for overdoing things. The Internet provides them opportunities to do so in one convenient place. In humans' distant past, life-and-death issues represented more physical than virtual threats. Our ancestors felt fortunate to be able to hunt, fish, and gather enough to eat and clothe themselves; find shelter in caves; build a fire for warmth; live long enough to reproduce; walk a few kilometers to see pretty vistas; and share stories in their very small and isolated communities. Contrast that with the "difficulties" of our more modern lives. Billions of people struggle with "life-and-death issues" of choice, such as—among others—what career to pursue; amenities to add to their homes; brand of cereal to eat for breakfast; clothes to wear; places to travel and how to get to them; media-based entertainment choices; and whom to interact with and when.

The prospect of being offline leaves many people feeling bereft. There is truth in the clichés that young people cannot imagine what life was like before the Internet and that baby boomers wonder how they ever got on with their lives back in the day without cellular phones.

Consumers are now faced with a myriad of choices of new technologies, particularly in the last quarter century. Whereas ancient peoples had simple diets, billions of people today (though by no means everyone) have a cornucopia of food choices. Whereas ancient peoples may have found that consuming some edible plants and herbs made them feel better or even high, today people in most countries have a pharmacopeia of natural and synthetic drugs available. Whereas ancient peoples could connect intimately with others, social networks were much smaller. People today have media resources to connect in real time with hundreds of millions of others (Facebook has 500+ million members), including a multitude of possible sex partners for trysts in either grounded or virtual reality.

One study in particular demonstrates changes in social networks. In 1969, social psychologist Stanley Milgram (famous and infamous for his "obedience" studies, in which participants were forced to deliver what they thought were severe or even lethal electric shocks to other people) reported that a chain of less than six people, or six "degrees of separation," connects any given individual with any other across the world.

Milgram randomly chose people from random locations in the United States, and asked them to "start" a letter-forwarding procedure to a target recipient in Massachusetts whom they did not know. Each "starter" received instructions to mail a folder via the U.S. Post Office to a recipient, but with some rules. Starters could only mail the folder to someone they actually knew personally on a first-name basis. When doing so, each starter instructed their recipient to mail the folder ahead to one of the latter's first-name acquaintances with the same instructions, with the hope that their acquaintance might by some chance know the target recipient.

Given that starters knew only the target recipient's name and address, they had a seemingly impossible task. Milgram monitored the progress of each chain via returned "tracer" postcards, which allowed him to track the progression of each letter. Surprisingly, he found that the very first folder reached the target in just four days and took only two intermediate acquaintances. Overall, Milgram reported that chains varied in length from two to ten intermediate acquaintances, with a median of five intermediaries (i.e., six degrees of separation) between the original sender and the destination recipient.

This concept became popularized by Jon Stewart's *Daily Show* in the mid-1990s, with the popular game—according to its creators, "a stupid party trick"—called Six Degrees of Kevin Bacon. The game was based on Milgram's experiment with a twist: that any actor can be linked through his or her film roles to actor Kevin Bacon within six degrees of separation. Over the past fifteen years, the Kevin Bacon game has become a board game, a book, a software package, and a cultural phenomenon. Even Bacon himself has become a fan of the concept, alluding to his centrality as a node in the vast Hollywood network in cameos, sitcoms, and television commercials.

There are so many direct and indirect contacts via the Internet today that scientists need supercomputers to even attempt to track who is connecting with whom directly or indirectly by connecting with someone who is networked to that person (one degree of separation), or by connecting with someone who is networked to someone else who is connected with that person (two degrees of separation), etc. So-called network science is, well, very complicated. Nevertheless, Microsoft researchers, after analyzing over 30 billion e-mail messages, concluded that Milgram's six degrees of separation formula was spot-on! Any two people in the world can be connected digitally to each other via an average of six intermediate steps.

We believe that as the number of people "wired" to the Internet grows, Milgram's magic number will decrease, resulting in more relationships that are less "separated" among the world's population. This presents a dilemma—more "friends" but less time for each.

Social networks are comprised of individuals (or organizations) called "nodes," which are tied (connected) by one or more specific types of interpersonal relationships, such as friendship, family, commerce, shared interests, etc. Sites that serve to facilitate social networking are widespread. The most visited Web sites are all somehow involved in social networking. These sites—for example, MySpace, Facebook, Bebo, Flixter, Adult Friend Finder, and Classmates—in 2010 included one billion registered users. In 2010, 2 billion people were connected to the Internet. We would not be surprised if by 2020 social network sites become ubiquitous: hundreds of thousands, if not more, will be available, most of which likely will be more specialized and involve far fewer users.

A thousand or more years ago (and even today, among scattered—often geographically isolated—groups and tribes), the small village was the site where people "networked" physically. Such hamlets provided a relatively small face-to-face social network. Even at the beginning of the twentieth century, in the United States and many other countries, population was 60 percent rural and 40 percent urban. This ratio slowly reversed during the twentieth century—by 2000, the U.S. population was 80 percent urban and 20 percent rural.

Psychological distances among people throughout the world have become much shorter. For example, young readers may be unfamiliar with the idea of "long-distance" phone charges—to them, the concept of making a call that costs more as a function of mileage is foreign. Older readers sometimes become anxious when phoning distant relatives, as the association between long distance and phone charges is difficult to shake. Today, the ways people conduct voice communication are quite varied. Indeed, during the writing of this book, we communicated via landline, cell phone, Skype, e-mail, chat rooms, etc.

What was once considered outlandish has become normative. When President Clinton took office in 1992, there were only fifty or so Web sites. Today there are hundreds of millions, though the number is growing so fast it is impossible to estimate with any degree of accuracy.

Even at the peak of the dot-com boom in the late 1990s, a person who used a "computer dating service" was considered either desperate or brave (and often both). However, the tides have turned, and now, especially in places like San Francisco's Bay Area near Silicon Valley, online fee-based social-networking sites for in-person dating have become a big business. One might even argue that online dating has made the concept of "the singles bar" obsolete. Checking out an online profile before a date gets much of the work done beforehand. Height, weight, income, and other demographic traits are required by the online dating sites, so the mystery of what one's date does for a living is solved in advance. Most Web sites take this a step further, and allow users to express their personality, hobbies, and even their "turn-ons" and "turn-offs." While it is still far from an exact science, the guesswork that was inherent in the typical blind date before the digital age has been drastically reduced.

VIRTUAL-REALITY ADDICTION

Clearly, the proverbial "village" that raises a child these days can be quite large. It can also be quite dangerous.

Think about yourself or someone you know as you ponder these yes or no questions.

WARNING

INTERNET USE MAY BE HAZARDOUS TO YOUR HEALTH

Do you/they:

Spend hours online without a break?
Prefer to spend time on a computer over friends and family?
Lie about the amount of time spent online?
Hide what is done online?
Check e-mail several times an hour?
Hear family complaints about the amount of time spent online?
Constantly think about being online—even when offline?
Log on while at work or school *instead* of working or studying?
Have suicidal thoughts stemming from lack of control over online behavior?

If the answer to even a couple of these questions is "yes," you may have an addiction problem. If the answer to the last question is "yes," then immediately put down this book and see a therapist.

The concept of addiction has long been tied solely to substances.

People didn't worry about becoming addicts if they didn't drink or do drugs to excess. It may be the case that alcohol dependence/addiction has been a problem for exactly as long as humans have consumed alcoholic drinks. Sooner or later, when drugs, such as nicotine, cocaine, or heroin, become widely available, addiction becomes a problem. Until relatively recently, the criteria for "addiction" required the use of one or more actual substances and was defined primarily as a physiological dependence on them.

Today, fewer scholars and practitioners in the health-related disciplines believe that substances are a necessary prerequisite for addiction. One reason is that some people can ingest or inhale so-called addictive substances—nicotine, alcohol, marijuana, or even cocaine—without becoming addicts, which suggests that psychological dependency plays an important role in truly addictive behaviors.

Indeed, psychological dependency was not explicitly included in diagnoses of addiction until relatively recently. Modern definitions include compulsions to engage in non-substance-based activities that they know will be harmful. Interestingly, some neuroscience research suggests that the same reward centers of brain activity are associated with the development of both substance-based and non-substance-based addictions. Placing a bet for a gambling addict is neurophysiologically similar to lighting up a cigarette for a smoker.

The Internet serves as the delivery mechanism for many non-substance-based addictions, especially video games, sex, social networking, gambling, and others. Fortunately, there are solid data on possible Internet addiction rates in the United States. Elias Aboujaoude, a psychiatrist at Stanford University Medical School, and his research team conducted a survey on a nationally representative sample of more than 2,500 adults in the United States on problematic Internet use, more than half of whom responded. Their results indicated that:

- 13.7 percent found it hard to stay away from the Internet for several days at a time
- 12.4 percent stayed online longer than intended often or very often
- 12.3 percent had seen a need to cut back on their Internet use at some point
- 8.7 percent attempted to conceal nonessential Internet use from family, friends, and employers
- 8.2 percent used the Internet as a way to escape problems or relieve negative mood
- 5.9 percent felt their relationships suffered as a result of excessive Internet use

Conservatively speaking, these data mean that at least 30 million adults in the United States may be Internet-addicted. But what are they addicted to, exactly? The big four in no particular order appear to be games (thousands and thousands exist), sex addiction (including pornography, sexually oriented chat rooms, etc.), social networking, and gambling. Others include auctions (e.g., eBay), online rummage sales (e.g., Craigslist), and information sharing (e.g., blogging and tweeting). The big four produce many billions of dollars every year for publishers of online games, pornography, and gambling sites, and revenue is increasing substantially every year. We expect that as virtual reality saturates human life, these addictions will become an even greater, worldwide problem.

It took many decades for nicotine addiction to be accepted as a major global health problem. Analogous to the cigarette's efficiency as a nicotine delivery device, the Internet has become the most efficient delivery device for non-substance-based addictive material to date. Tens of millions of Web sites support Internet-based addictions. It is not surprising that face-to-face twelve-step programs akin to Alcoholics Anonymous exist for Internet addicts. In fact, there are online twelve-step programs.

For example, South Korea leads the world in Internet-based addiction recovery programs, having recognized a major problem among their youth. A PBS *Frontline* documentary titled *Digital Nation* powerfully describes the experience of attending a South Korean boot camp to cure Internet addiction. In a typical camp, a group of young teenagers are forced from their homes into a group home, where they attend workshops, support groups, and engage in other forms of treatment. One of the most potent tools used by the boot camps is simply forcing the children to go outside—for example, pitching tents, sleeping outdoors, and even doing lawn work. Sometimes these boot camps are effective, other times not. The segment on PBS ends with a young boy shrugging when asked if he has changed as a result of the boot camp. He clearly is dreaming of his return to the online world, and says as much to the host.

Of course, Internet addiction is not limited to Southeast Asia. Rather, South Korea is arguably the "canary in the coal mine"—an early warning of worldwide addiction problems. It shouldn't come as a surprise that addictions occur in the United States. On October 15, 2009, *USA Today* described a female marketing contractor hopelessly addicted to the game *FarmVille* on Facebook. She can't sleep at night due to worry over her virtual crops and she uses it to escape from city life. She is not alone. On October 27, 2010, a mother in Florida pleaded guilty to second-degree murder for shaking her three-month-old infant, ultimately causing his death, because his crying interrupted her *FarmVille* play. *FarmVille* is one of the fastest-growing games ever, with more than 80 million players in 2010.

The first dedicated Internet addiction treatment facility in the United States, reSTART, opened in 2009 in Fall City, Washington. Fittingly, it is located just a few miles from Microsoft's headquarters in Redmond. Like drugs, alcohol, and food, virtual-reality sites can affect people so strongly they surf compulsively. The consequences are dire to the addicts and their loved ones. People flunk out of school, lose jobs, break up families, cheat their employers, and have died.

Yes, people even die! We hope that clause did not slip past you. On August 10, 2005, BBC News reported that a South Korean man died after playing almost nonstop for fifty hours, collapsing of exhaustion in an Internet café. On May 27, 2010, the *Daily Telegraph* reported that a fanatic French video-gamer hunted down and stabbed a rival who had killed his war-game avatar. The attack occurred not in the confines of the virtual game but on the victim's physical doorstep. The Frenchman knocked on his rival's door and stabbed him. On October 13, 2009, *PC News* reported that, in a fit of frustration, a Swedish teenage boy allegedly stabbed a girl after an Internet glitch left him unable to play *StarCraft* online.

SUCH NEWS STORIES TO THE CONTRARY, IT IS ALSO THE CASE that virtual reality generally lessens people's anxieties, at least temporarily. Dreaming, drinking a few beers, or smoking a joint help humans escape their worries. The Internet can become a "monkey on people's backs" because it provides a quick and easy escape from daily life.

But there is another reason that the Internet is so compelling. Humans are gregarious social beings who like to be with other people. Loneliness leads to all sorts of problems, psychological and otherwise, as Kipling Williams's work on ostracism (described in chapter 4) has demonstrated. Data from some scholars investigating social interaction on the Internet support this view.

More than a decade ago, Katelyn McKenna, now at the Interdisciplinary Center for Internet Psychology just outside of Tel Aviv, and John Bargh, social psychologist at Yale University, published a prescient paper titled "Causes and Consequences of Social Interaction on the Internet." They maintain that people use the Internet to meet fundamental human needs, including belonging and self-esteem, arguing that the Internet is no different than a pot luck dinner in terms of fulfilling the need to be social. McKenna and Bargh found that people generally suffer *less* loneliness as a result of Internet use. Lyrics from Brad Paisley's country-western song "Online" about a lonely, not-so-attractive man who lives at home with his mom illustrate the point. On the Internet, Paisley claims, "I'm out in Hollywood. I'm six foot five and I look damn good . . ."

On the other hand, in a research study that began in the mid-1990s (likely the first longitudinal study of Internet use), Robert Kraut, a psychologist at Carnegie Mellon University, led a team who studied people "before and after" they used the Internet. They found that greater use of the Internet was associated with declines in participants' communication with household family members, declines in the size of their social circle, and increases in their depression and loneliness.

Regardless of who is correct in the loneliness debate, other consequences of addiction can be severe and troubling. For example, as reported by the Associated Press in 2007, a couple in Nevada were convicted of child neglect. They were so wrapped up in their Internet gaming addiction that they left their children to "fend for themselves." The children had infections, were malnourished, and one child even had cat urine in her hair. More troubling is the fact that the couple actually had inherited fifty thousand dollars. They chose to spend it on cutting-edge equipment for gaming instead of feeding their offspring. When discovered, they faced criminal charges and up to a dozen years in prison.

In the realm of substance abuse, some argue that addicts are not only dependent on substances, e.g., alcohol, nicotine, heroin, etc., themselves but also on the delivery devices they use, such as beer bottles, cigarettes, syringes, etc. It is not far-fetched to consider that people who have Internet-based addictions become strongly attracted by Internet delivery devices—for example, smart phones and computers—even when it is physically dangerous to use them, say, while driving. However, whereas one can stay out of bars, tobacco stores, and crack houses, people are not likely to abandon digital media, as modern daily life typically depends on computers and smart phones. Hence, the Internet addictions may be particularly difficult to shake.

VIRTUAL-REALITY ADDICTIONS

According to *Forbes* magazine, 5 to 10 percent of Internet users compulsively access Internet applications and, arguably, can be said to be addicted. If Internet addictions are strong, addictions stemming from *immersive* virtual-reality experiences should be even stronger. The experience of using a joystick to traverse a narrow bridge over a virtual crevasse via a 2-D display is nothing compared to walking over that bridge while one's movements are tracked and the scene is rendered to multiple senses in 3-D. Over the last few decades, near-totally immersive systems have been used primarily by the military and research institutions. But as platforms such as Wii and Kinect proliferate, and as stereoscopic displays become affordable, consumers will have more immersive 3-D experiences, resulting in much stronger virtual-reality addictions.

Let's briefly consider what more immersive technology will do in terms of the "big four" Internet addictions discussed above: social networking, gambling, gaming, and sex.

Social Networking

"Classic" social-networking sites involve a blend of text, audio, photographs, and video. New platforms, such as *Second Life*, allow people to create their own avatars, meet others, and interact with them. They also buy property, build houses, and choose furnishings. In fact, a whole economy based on "Linden dollars," which have an actual U.S. dollar exchange value, has evolved in *Second Life*. Some users spend more time on *Second Life* than on just about any other activity outside of sleeping or working.

Can *Second Life* be addictive? Many individuals testify that it is. Such compulsive *Second Life* users seek out treatment such as OLGA (Online Gamers Anonymous). There are even addiction treatment centers in *Second Life* for addictions like alcohol and tobacco. Of course, an online addiction treatment center actually located *in Second Life* would create quite a paradox for the sufferers.

But even *Second Life* does not represent the ultimate example of social networking addiction. Instead of using a keyboard, joystick, Wii, Move, or Kinect-based device to navigate a future version of *Second Life* or Facebook, imagine donning a headset and beaming yourself into a 3-D immersive social-networking site, meeting people, building a home, going to clubs, engaging in almost every activity known in grounded reality. If avatar movements are generated by one's corresponding physical movements, and if one's perceptions are expanded beyond sights and sounds to also include touches and smells, then one's "second" life will be a whole lot more like their "first" one.

Given that everyone's movements can be tracked, rendered, saved, and replayed in virtual reality, one can relive an experience or even "change the past." Had a great virtual tryst? Play it again. Interrupted? Pause and continue later. Turns out the tryst wasn't all you thought it would be? Hit the "wipe out" key and the experience is gone—kind of like taking the red pill in *The Matrix*.

Consider what may be the second-most-popular activity in today's *Second Life:* virtual dancing in clubs. (We will talk in detail about the most popular one later in this section.) *Second Life* has hundreds of virtual dance clubs with the look and feel of physical dance clubs, including features like strobe lights, fancy clothes, cocktails, and hip music. In any given club, at any given time of the day, visitors can view dozens of avatars dancing. But what is virtual dancing? A user clicks a button on their computer and the avatar subsequently performs a dance routine, each one typically lasting about thirty seconds—kind of a *Dancing with the Stars* on automatic pilot via avatars.

Why is virtual dancing so popular? We recently had lunch with the chief technical officer of *Second Life*, and he told us that none of the creators had any idea that dancing would be such a popular activity. Indeed, the "dance" experience for the user is hitting a button once every thirty seconds. But, for some reason, it has caught on. Imagine how much more popular networked dancing will be once it involves users' physical movements. Indeed, one of the most popular games in video arcades has been *Dance Dance Revolution*, a game in which players score points according to how well they time their dance moves (tracked by sensors on a physical dance platform) to the patterns presented to them on the game monitor.

Gambling

Reports indicate that online gambling sites are increasing and likely to overtake offline wagering in the glitzy casinos of physical reality. Gambling addicts need not travel to Vegas or Monte Carlo to gamble or even to go to church to play bingo. They can just sit home and bet on sports, races, and elections, or play poker and other casino games without leaving their easy chairs. And they can lose money just as easily as in "real" casinos. People like to gamble.

Young adults constitute a disproportionately large segment of online gamblers. Attracted by the popularity of poker tournaments on television, a high percentage of college men gamble online. "I think it's a devastating illness, it's an illness that if it's not treated, will end up that the person's whole lifestyle will be affected," says Ed Looney, executive director of Council on Compulsive Gambling. Nevertheless, the online gambling industry rakes in mind-boggling amounts of cash. Given how pervasive gambling is, we can only conclude that wagering behavior strikes a very resonant chord in the human psyche. People find all sorts of ways to bet, whether legal or not.

Recall the virtual reality casino described in chapter 5, an immersive one complete with slot machines, a blackjack table, a dealer (an agent), cards, chips, etc. Even in an early version, research participants reported being quite immersed. However, they also pointed out that except for the voice of the blackjack dealer, our casino was eerily quiet. They wanted casino sounds! We obliged, and the casino became even more immersive. Indeed, lab visitors, as well as research participants, tell us they enjoy sitting at the virtual blackjack table, counting their chips, watching each other play, and yelling "Yes!" or "%$#&," depending on the outcome of a particular hand. People are willing to play for hours.

Will virtual-reality casinos increase gambling addiction? Physical casinos have multiplied throughout the world, especially in the United States—people crave "ambience." Treating oneself to a stay at a casino is much more than a mere gambling experience. It's a chance to enjoy the glitz of exotic worlds, which is why Las Vegas casinos create historically themed places to play—Ancient Rome, the Pyramids, New York, Venice, and Paris are just a few examples. Patrons come to see celebrities and rub shoulders with beautiful people. Just as our research participants wanted to hear casino noises in our virtual-reality casino, people love the excitement of the whole scene. Why do casino owners go to such great lengths? Because they know that their "glitzy" environments increase their revenues.

Virtual reality takes the concept of glitz to new levels. While Vegas casino environments have physical limits, virtual ones can present any scene imaginable. One could gamble in the Louvre, underwater in the Great Barrier Reef, even on the moon. One could sit down to play blackjack with their favorite movie star as the dealer. Casinos will be stocked with audiences who cheer at players' victories, and gorgeous men and women who cling to their sides. Casino owners are rumored now to employ devious strategies to keep gamblers at the tables, ranging from designing floor layouts to hide exits to pumping oxygen in the room to keep people awake. Given an arsenal of virtual tools, casino owners might be able to build virtual havens gamblers will never want to leave!

Virtual Game Addiction

Games have provided ways to interact in virtual worlds for as long as any of us can remember. Traditional board games—*Monopoly*, *Risk*, and even *Chutes and Ladders*—draw people's minds away from their worries. Technologically, board games gave way to home console and computer video games, such as *Pong*, *Tetris*, *Pac-Man*, sports-based simulation games, and then on to online interactive games: e.g., *The Sims*, *World of WarCraft*, and *EverQuest*. In less than a year, the game *FarmVille* went from 350 or so users on its debut day to more than 80 million users.

Do actions in these games affect life in the physical world? A substantial scientific literature describes the link between game violence and physical violence. Brad Bushman, an expert on media violence at Ohio State University, argues that fifty years of scientific work on video games demonstrates that "exposure to media violence causes children to behave more aggressively and affects them as adults years later." Taking a minority but countervailing view, Jonathan Freedman of the University of Toronto maintains that there is no evidence to support negative effects of media violence.

The critical question that relates to addiction is how games will change once they become housed in virtual reality. To begin to answer this question, Susan Persky, a research scientist at the National Institutes of Health, developed a "shoot-'em-up" game to test the relative effects of 3-D immersive virtual reality, compared to a typical desktop video game, on postgame hostility. She found that, compared to the desktop condition, participants in the immersive scenario not only reported more feelings of aggressiveness but their hearts pounded harder and faster and their blood pressure rose during the game. They exhibited strong effects from being immersed in the violent game experience.

Sex

Unsurprisingly, humans are susceptible to seduction, whether in physical or virtual reality. Throughout human history, virtual depictions of attractive people, whether the product of internal mental processes like dreams and daydreams or the product of media technology (stories, painting, sculpture, audio, etc.), have been good business for their creators. The Internet has made such depictions so widely available that it is impossible to catalog them all—though many Internet sex addicts seem to be trying.

Consider the strange case of David Pollard and Amy Taylor, a British couple who met online and married offline. Though apparently faithful in physical reality, David admittedly still had sexual relationships in *Second Life*. The *Globe and Mail* reported on November 15, 2008, "After catching her husband with his virtual pants down . . . , Amy Taylor decided enough was enough." Had David been unfaithful? Supporting his claim that he had not, he argued that cybersex is merely a shared fantasy and that only his avatar had developed a virtual sexual relationship. Apparently, David's avatar was a "real" enough alter ego for his wife.

In *Second Life*, like online dancing behaviors, online sexual behaviors are somewhat strange. Avatars, often wearing biological appendages that have been purchased at virtual stores, virtually copulate as the users hit buttons to activate animation sequences on their home computers. Some reading that last sentence most likely find it hard to imagine how this can be a fulfilling experience. But it turns out that, for many, sex in *Second Life* is a worthwhile endeavor. In the class we taught at Stanford, eighty students spent hours each week inside *Second Life*. They were instructed to explore and catalog the types of behavior they observed. We were amazed by the amount of sexual activity that our students reported. The frequency was startlingly high, close to one-quarter of all activities. Of course, this number is relatively "unscientific," as the students were estimating the frequencies based on their experiences, but nonetheless, they assured us that the amount of sex that goes on in *Second Life* is startling.

As virtual reality becomes more immersive, virtual sex will become more and more . . . satisfying. Indeed, "teledildonics" is an emerging field that incorporates haptic devices, those capable of transmitting virtual touch, into virtual sexual experiences. When one thinks about the ability to "archive" certain people (recall, in chapter 9 we discussed how to construct 3-D models of specific people), and then the ability to animate those models with haptic devices designed for sexual pleasure, the potential for even stronger addiction becomes scary.

To paraphrase Dennis Miller, the day a teenage boy can buy an avatar that looks like a supermodel for $19.95, virtual reality is going to make crack look like a cup of weak, instant, decaffeinated coffee.

WHERE NEXT?

To be sure, not everyone who uses the Internet is addicted. Not everyone who enters social networks becomes addicted to them, just like not everyone who smokes marijuana becomes addicted. And some may argue that all humans are "addicted" to socializing because that is what humans have evolved to do. Nobody knows any absolute hermits. As we have emphasized previously, people are naturally gregarious beings motivated to be with others and not wanting to be alone. The Internet and virtual realities easily satisfy such social needs and drives—sometimes so satisfying that addicted users will withdraw physically from society.

CHAPTER TWELVE

VIRTUALLY USEFUL

VISITORS TO OUR LABS FOR VIRTUAL-REALITY TOURS HAVE A REmarkably predictable reaction: "This is really cool! . . . But what's the point? How does it affect my life?" We answer, "It has for some time, does now, will even more so in the future."

Although virtual-reality applications may not make the eleven o'clock news every night, this technology has been changing how people design products, defend nations, educate children, provide medical care, travel, and create entertainment. Head-mounted displays and other virtual reality devices are not yet found in every family room. But think about the computer. Once upon a time, computers were only housed in large corporations and government organizations, but today personal computers appear in most American homes. Large organizations such as corporations, governments, and universities are currently the vanguard for virtual-reality technology, just as they were for computers decades ago. Byron Reeves (recall, in chapter 1 we described his work demonstrating that virtual experiences can be real) convincingly argues in his book *Total Engagement* that avatars will change the manner in which corporations and other organizations operate. Here, we examine just how this change will occur.

START YOUR ENGINES!

Time magazine ranked the fifty worst cars of all time in their fiftieth-anniversary issue. The 1971 Ford Pinto was in the middle of the pack, ranked twenty-second. A relative success in terms of fuel efficiency and compact size, the Pinto contained a fatal design flaw so notorious it became the quintessential example of Detroit's shortsightedness. Ford's designers positioned the gas tank underneath the rear axle, not uncommon for many cars, but unfortunately the back of the Pinto was so poorly reinforced that the tanks often exploded into flames during rear-end collisions.

Despite Ralph Nader's famous warning to the auto industry in his book *Unsafe at Any Speed*, published six years before the Pinto appeared in dealer showrooms, Ford simply did not perform proper safety testing on the car. As a result, Pinto drivers reportedly suffered horrendous injuries and fatalities. While there is disagreement among experts as to exactly how many deaths were directly attributable to this defect, many estimate the toll to be around twenty-five with as many severe burns.

When investigators discovered from internal corporate memos that company executives had been aware of the problem, the public relations fallout for the Ford Motor Company was severe. Corporate managers had estimated that the likely cost of lawsuits based on injury and loss of life due to the gas tank flaw would be cheaper than a recall to fix the design problem. Ford was clearly remiss in not safety-testing the Pinto properly. Perhaps their safety engineers did not want to risk the danger and expense of explosions in their facilities.

But at least Ford learned a lesson. By turning to virtual reality, they now not only more fully test their new car designs but do so safely and inexpensively.

Elizabeth Baron manages the Ford Virtual Reality Center, an arm of Ford's product development division. This center addresses the unique challenges of automotive design and engineering. In her keynote address at the premier virtual-reality engineering conference, IEEE VR, in 2009, Baron described several successful—and cost-effective—virtual-reality projects that have led to important automobile design changes.

In general, changing the size and/or shape of a physical automobile prototype, even "simply" (for example, to make the inside of the car roof 10 percent higher), can cost millions of dollars and consume lots of time. However, if a digital model is built, changing its height, width, length, color, etc., is relatively cheap and extremely quick. By investing in Baron's lab, Ford is able to efficiently optimize designs that work best for drivers. The vice president of engineering at Ford proclaimed that the virtual-reality lab saves six months in product development time, as well as millions of dollars.

Consumers want a car that fits them comfortably, but given the range of body types—take drivers with really long or short legs—it's no simple task to achieve. In the auto industry, the art of fitting cars to people is called "human scaling." Resizing an automobile involves determining whether designs work for people of different sizes, shapes, and preferences. Human scaling requires specifications such as interior size, seat ratios, and the placement of the steering wheel, brake and gas pedals, and dashboard controls are worked out by trial and error. To illustrate, Baron's group determined whether a particular sun visor would successfully block glare for drivers who differed in height. They built virtual-reality car models and driving environments to determine if the proposed sun visor would work, for example, for a four-foot-eleven female driving under a variety of weather conditions, including driving into the sun during various times of day. Baron was able to use virtual reality to control a factor that nobody can manipulate in the physical world—the position of the sun.

Baron's lab also used virtual reality to test how drivers of various ages responded to car prototypes. Their test results guided car designs that make entry and exit easier for older drivers. The AARP gave Ford an award for building automobiles that were well suited to the demands of older people.

Additionally, Ford wanted to understand how the shape of windshields affects driving distractions: for example, the way a driver's vision becomes diverted by attention-grabbing views outside the car. Using virtual reality, Baron's team was able to design windshields that minimized distraction and improved driving performance.

Because virtual-reality testing does not produce actual physical damage, it allows Ford to run experiments that would not be possible otherwise, for example, testing a life-saving system that detects when one's car is drifting into another lane. To actually cause drivers to drift into oncoming traffic obviously presents safety risks, but in virtual reality, scientists can capture "naturalistic" driving behaviors without risk.

Decision-makers at all corporate levels at Ford have embraced virtual testing as a crucial tool to produce safe cars, which keeps them economically competitive in the global marketplace. Indeed, according to Baron, during the recession of 2008–2009, the virtual-reality lab at Ford (the lone major American carmaker that avoided government takeover) was one of the few areas that didn't face cutbacks in personnel, speaking to how valuable virtual-reality technology can be, even to established manufacturing giants.

BUY ME!

In the United States, a major event occurs every winter—Super Bowl Sunday. In past decades, Super Bowl television advertisements cued viewers to get out of their seats, refresh the popcorn bowl, grab a couple of beers, and use their toilets—collectively creating the largest simultaneous stress on water and sewer systems in the country. Nowadays, Super Bowl television advertisements freeze conversation, enticing viewers to watch commercials, thereby reducing the danger to community sanitary systems. Indeed, a substantial percentage of viewers now care more about the commercials than they do about the game. Amazingly, a recent poll by Nielsen indicated that 51 percent of Americans reported enjoying the commercials more than the game itself!

Such an evolution in attitudes toward television ads is not limited to Super Bowl Sundays. In the past, companies spent only a fraction of their annual budget on advertising. For example, in 1960, Volkswagen spent approximately $800,000 on advertising. In 2005, they spent more than $350 million on advertising—more than four hundred times the 1960 amount! Market research accounts annually for billions of dollars in corporate spending. Somewhat analogous to Ford's need to pretest car designs, companies need to pretest logos, packaging, and advertisements in order to best gauge their effectiveness.

"Pick me!" is the simple goal of marketing—drawing consumers to products. The aim of market research is to determine how best to do so. One research method involves "focus groups," asking people to discuss why they buy products. As behavioral scientists,

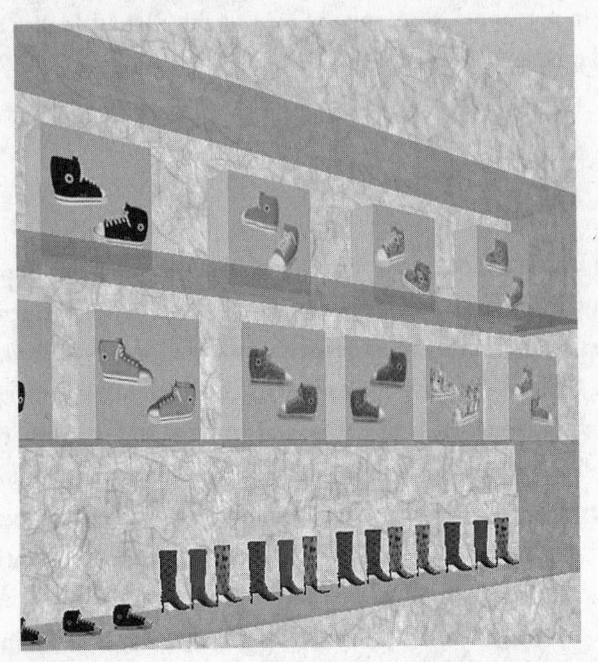

Shopping without dropping: A virtual-reality "test" store.

we find this technique less than optimal. The problem is twofold. First, the people comprising focus groups are typically not a perfect snapshot of the average consumer. Instead focus groups are often composed of people who supplement their income by attending such meetings regularly. Second, the "data" gathered from focus groups are very "fuzzy," because it is difficult to interpret these subjective, open-ended conversations.

In contrast, the perfect market research technique uses physical "test" stores, in which companies run experiments on large and demographically diverse samples of target consumer groups. Unfortunately, building test stores, filling them with product, and recruiting a random sample of shoppers is very expensive and, unsurprisingly, rarely used. However, virtual reality allows companies to conduct market research studies on large, representative, and diverse groups of likely buyers at a much more reasonable cost.

When a clothing manufacturer tests consumers' reactions to new shirt designs, they usually experiment with color, sleeve length, buttons, etc., and gauge appeal to potential customers. Virtual reality facilitates both design and testing. Just as is the case with a digital music file, once that virtual shirt has been modeled, it can be replicated hundreds or thousands of times in less than a second. To get the look and feel of a fully stocked store, a designer can literally hit the "copy" and "paste" buttons and the store is full of shirts!

The marketing team could then easily produce the stock for an entire virtual rack for the cost of digitally modeling a single shirt prototype—a few hours' work. Shirts often look different in carefully designed store displays than they do when viewed in isolation. Given that it is critical to examine consumer behavior in context, virtual-reality stores allow market researchers to capture shopping behavior in a "natural" environment. Even though there are some differences between virtual and physical stores, the lessons learned in virtual reality can be used to determine how the shirts, racks, and layout will appear in the physical stores. This "test virtual, build physical" technique also works for intangible aspects of store design such as lighting, aisle size and design, placement of mirrors and partitions, etc.

One day, immersive virtual stores will become functional 3-D stores in which consumers will shop "in person" as the necessary technology becomes common in people's homes. In fact, such stores and products can actually be individually tailored for every consumer, as is done today by online booksellers and movie rental companies, like Amazon or Netflix, which "learn" consumer preferences and make educated suggestions for buys and downloads. In terms of market data collection, virtual reality can be used to capture all of a consumer's movements in an immersive virtual store: where the consumer walks and, important, where she is looking and what she is touching and purchasing. Using virtual technology, market researchers can mine these valuable behavioral datasets to test their ideas or hypotheses about packaging, store layouts, and advertisements.

This kind of research gathering is not necessarily reserved for the future. Currently, many corporations build virtual simulations to determine consumer preferences. For example, the Fortune 500 corporation Kimberly-Clark designs and tests product placement, colors, graphics, layouts, and displays for toiletries and other products in virtual stores. While consumers shop, their eye movements and other motions are tracked. Findings from Kimberly-Clark's studies have been used to optimize product placement in grocery stores such as Safeway, resulting in increases in sales in "all toiletries tested." The research and development team from Kimberly-Clark has also visited our lab at Stanford to explore ways to increase immersion in their simulations even more.

Virtual reality solves another problem for market researchers. Currently, when a researcher wants to market a product to consumers—for example, a new vacuum cleaner—she goes to the local mall, pesters people who appear to fit the demographic (based on their age, gender, and perhaps waistline) in which she is interested, and tries to find ten or so such people in order to conduct a focus group to discuss various marketing ideas pertaining to the product. This can take hours or even days, depending on who happens to be wandering around the mall. On the other hand, because virtual reality can reach anyone online, the population available for sampling is much more closely representative of the actual population of interest and can be sampled randomly. This ultimately results in statistics that are more accurate than the "mall-trolling" technique.

In another application, marketers are rapidly adopting virtual reality for product placement. Within traditional media, we have seen the practice move from obscure to obvious. Consider how, decades ago, placement was subtle. For example, in the 1960s sitcom *Mr. Ed*, one can see many images of Studebaker cars placed prominently in various episodes. Not surprisingly, Studebaker was a primary sponsor of the show for a number of years. Contrast that with more recent entertainment that wraps entire plots around particular products or purveyors. For example, in the popular television show *The Office*, a sitcom about a small paper company called Dunder Mifflin, there is a great deal of discussion about their biggest competitor—Staples. Actual products from Staples become part of the plot. For example, in one episode, the boss uses a Staples paper shredder to make a salad, and in another when a Dunder Mifflin employee gets fired, he chooses a job at their competitor: Staples. The Staples Corporation pays for this product placement.

Video games utilize similar strategies. For example, Castrol Syntec paid to have their own, clearly branded quick-lube shops in Electronic Arts' *Need for Speed*, a racing game in which players need to maintain their race cars. Similarly, in *CSI: 3 Dimensions of Murder*, a game in which players solve crimes, Visa pays to include their fraud-protection service, which alerts players to a clue—a stolen credit card. As virtual reality becomes more immersive, the competition for product placement will undoubtedly become more intense. The lingering effect of such placements will become more compelling as consumers become able to see products in 3-D, touch them via haptic devices, and even smell them using virtual olfactory senses.

CSI: VR

In an episode of the popular dramatic television program *CSI: NY*, law enforcement officials identified (and eventually arrested) a homicide suspect using virtual reality. The perpetrator murdered people in physical reality, but tracked his victims in the online virtual world *Second Life*. That "fact" enabled the investigators to track and find him by uncovering clues in the virtual world and tying them to the user's physical identity.

Virtual reality has been used in the U.S. criminal-justice system, though admittedly in a less action-packed way than on the *CSI* franchise. One of the most promising courtroom applications of virtual reality is re-creating crime or accident scenes. Such scenes naturally include inanimate objects, witnesses, victims, and suspects. They also can include varying ambient conditions—e.g., lighting, temperature—and even weather conditions such as rain or fog, background noise (traffic sounds), and anything else that may have occurred during the crime itself.

In court, such re-creations can help to verify or impeach the testimony of witnesses, test forensic theories, and enhance courtroom participants' understanding of a past event. Because it is possible to render the 3-D scene to a third-party observer from any point of view, virtual reality can be used to demonstrate whether a witness actually could have seen the crime.

Examine the screen shot on the next page, of a virtual world depicting doctors and nurses gathered around an anesthetized patient undergoing open-heart surgery—a scene we created nearly a decade ago for a mock trial conducted in the Courtroom of the 21st Century at the College of William and Mary Law School.

In this hypothetical case, the patient died following surgery to implant an aortic stent—a coiled, springlike device in his main blood vessel, the aorta—to keep this critical artery open. After investigating, the Food and Drug Administration charged the company that made the stent with criminal negligence for coating the stent with an unapproved pharmaceutical. The company's defense

A scene from the hypothetical surgical case we created for a mock trial at the College of William and Mary Law School.

argued that the patient died because the surgeon inserted the stent backward. One of the nurses testified that he saw the surgeon insert the stent backward. Another testified that she saw the surgeon insert it correctly.

Everyone involved in the mock trial agreed that the virtual scene was spatially accurate and that it could be used to determine what exactly any of the witnesses in the room could actually have seen. We inserted a *view frustum* (i.e., the white mesh structure) into it. If a witness's eyes were never inside this frustum, they could not have seen the surgeon's hands within the chest cavity. Consequently, lawyers in the mock trial were able to use this simulation to discredit the testimony of both nurses, whose heads were never situated within the frustum.

Lawyers and judges can also use virtual reality to enable jury members to experience a crime or accident scene "firsthand." Unlike a two-dimensional map, chart, or another audiovisual aid, such as a photograph, jurors can enter a re-creation in virtual reality. Jurors can experience the visual perspective of any person who was actually at the scene. Furthermore, they can move around and experience the environment from any point of view.

We have worked with the Federal Judicial Center to help them use virtual reality to study how reconstructing crime scenes digitally can improve eyewitness identification via "police lineups." Compared to witnesses looking at suspects' photographs, which is how approximately 75 percent of police lineups occur in the United States, a virtual sequence of suspects embedded in a re-creation of the crime scene itself can improve eyewitness accuracy.

A witness wearing a head-mounted display and walking among the avatars can see them from novel angles. Moreover, she is able to get extremely close to their heads and faces—something actual witnesses are not only prevented from doing but are often afraid to do. If the crime occurred in a liquor store, the witness could view the lineup in a virtual re-creation of the liquor store without ever having to physically return to the scene of the crime. This is particularly useful for witnesses who would be traumatized by a return to the actual crime scene, or when the crime scene no longer exists (as is often the case with arson, for instance).

To examine such contextual cues, we ran an experiment that "locked down" the viewpoint of a witness toward the suspect. In other words, if a witness only saw a suspect from a fifth-story balcony, then it makes no sense to show him standard mug shots of a "straight-on" view. Indeed, a bird's-eye view that takes into account the correct visual angle would be more appropriate. We paid participants to come to the lab, where they witnessed a "faked"

A virtual lineup can improve eyewitness accuracy.

crime. One of our research assistants played the unfortunate role of a criminal. He pretended to be a participant waiting for the experiment to begin and, after a few minutes, loudly proclaimed that he had to be somewhere else and couldn't wait for the stupid experimenters. He not-so-subtly grabbed the cash box from the lab as he sprinted out of the room. The other participants were flabbergasted, and all of them thought they were witnessing a real crime. One participant even chased the "thief" down the hallway, yelling for him to stop.

This procedure ensured that participants were aroused during the crime, as memory works in different ways, depending on one's physiological responsiveness. We then gave the participants a partial debriefing, and told them it was not a real crime, but that we were studying memory. The participants, acting as witnesses, then viewed a virtual lineup via a head-mounted display, and saw a lineup of virtual humans—one of whom was modeled after the confederate who stole the cash box, while the other six bore some resemblance to him. There were two conditions. In one, participants saw all lineup members from the same distance—which is how a series of mug shots work, based on the distance from the camera to the face. In the other, they saw the suspects from the approximate distance that they had seen the thief in physical space while sitting next to him in the waiting room. They then had to identify which virtual human represented the person who stole the cash box. Our data demonstrated that, when we matched the viewing distance to fit the crime context, accuracy improved. In other words, it was counterproductive to give witnesses extra information about a suspect's face that they could not have encoded during the actual crime.

About 75,000 people per year become criminal defendants due to eyewitness testimony. Each year, especially now that DNA evidence is gaining momentum, the justice system discovers people wrongly incarcerated because of overconfident and inaccurate witnesses. A recent estimate made by a group called the Innocence Project shows that of the 232 people who were exonerated by DNA up until 2008, three-quarters had been convicted on the basis of at least one witness misidentification.

In sum, the legal system currently uses simple animations and physical models of crime scenes during trials. As technology becomes more immersive, the uses of virtual reality will become pervasive. Given this potential, it is especially crucial for the courts to ensure that the use of virtual reality is regulated properly, so that the technology aids the search for truth rather than provides unnecessary "bells and whistles" to woo a jury.

MEDICINE AND THE VR ER

Health care has driven much virtual-reality technology because of its potential cost-effectiveness. Hence, one of the most successful virtual-reality conferences is Medicine Meets Virtual Reality, which brings together behavioral scientists, computer scientists, psychiatrists, medical doctors, and neuroscientists from academia and the professional world. The goal of the conference is to develop uses of virtual reality to improve health care in various areas.

Surgical Training

Perhaps the earliest and most used virtual-reality medical application is surgical training. Many surgical devices are difficult to master, for example those that are designed to be used around bones or organs. Moreover, the traditional props used in surgical training—human cadavers—are expensive, difficult to obtain, and can't be reused. Alternatively, virtual-reality displays of the human body, coupled with haptic devices that provide resistance to touch, are an excellent, cost-effective way to train surgeons.

Robert O'Toole is a practicing physician who has evaluated surgical simulations at a number of universities, including Harvard and the University of Maryland. In a landmark study, he and his team trained twenty or so experienced surgeons and third-year medical students using virtual reality over the course of ten sessions. By measuring patient "tissue damage," they demonstrated that medical students as well as seasoned surgeons improved their skills over time as a result of virtual training.

Another medical application involves the construction of 3-D models of bodies and tumors to guide surgeons conducting minimally invasive procedures, such as laparoscopic surgery, in which small holes are punctured through the skin and thin instruments inserted in order, for example, to excise a tumor. Radiologists can use imaging techniques such as ultrasound or magnetic resonance imaging to develop 3-D models of the "hidden" tumors prior to surgery and "implant" them in a virtual model of the patient's body. With the physical patient anesthetized and immobile, the surgeon can don a head-mounted display to "see" the organ they are operating on underneath the skin. Furthermore, tracking technology allows surgeons to monitor the position and angle of the surgical instruments inside the body. As a result of this advance, surgeons increase success rates while minimizing complications associated with more invasive surgical techniques. With robotic arms added, surgeons can even do a surgery from the other side of the physical world.

Physical Rehabilitation

People who have injured themselves and undergone strenuous physical therapy know how difficult and tedious the process can be—driving to the clinic twice per week, repeating the same painful motions over and over again. Doctors have long known that the key to successful healing is adhering consistently to the therapy regimen. But people often have trouble being "good patients," given how unpleasant the experience can be. Consequently, a number of physicians and scientists have developed virtual reality therapy techniques.

Virtual reality has a number of unique advantages for physical therapy. First, and most important, it can make a repetitive and tedious process interesting. For example, consider the technique utilized by Judith Deutsch, a professor who has evaluated the use of virtual reality for physical therapy at a number of universities, including Stanford, the University of Southern California, and New York University. She and her colleagues developed a system to aid those rehabilitating injured ankles.

The patient uses his foot, which is tracked, to move a virtual airplane through sequences of square windows that are designed to require therapeutic foot movements. The advantages of such a system are numerous. First, studies have shown that the patient is more likely to complete the exercises. Second, unlike a preset list of repetitions to follow, the sequence of windows can be artificially intelligent. In other words, the system can analyze the movements of the patient in real time, and determine, for example, if the patient is fatigued. If so, then the system can adjust, and operate at a more appropriate rate. Finally, the system can provide movement constraints, that is, use visual, auditory, and haptic feedback to ensure certain movements are avoided. In this sense, the occurrence of reinjury, which is always a possibility in therapy, can be reduced.

Pain Management

Physical pain is subjective. People differ in the amount of pain they suffer from the same exact injury, and people sometimes feel pain when no related injury can be identified. For example, when Admiral Lord Horatio Nelson lost his arm in the battle of Tenerife in 1797, he later felt "phantom-limb" sensations—experiencing direct pain from his fingertips pressing into his arm, even though none were. Lord Nelson interpreted this sensation as proof that the soul could survive physical obliteration.

More recently, scientists have used virtual reality to manage the subjective experience of pain. At the University of Washington, psychologist Hunter Hoffman first used an immersive virtual winter scene to counter severe physical pain. Patients with third-degree burns donned a head-mounted display and floated through a snow-covered canyon filled with penguins, snowmen, igloos, and woolly mammoths. Meanwhile, doctors removed dead skin from their burn areas.

A world without pain: Doctors use a simulated winter scene to counteract severe pain.

The virtual therapy was so effective that reported pain was reduced by 50 to 90 percent in clinical trials. Interestingly, it turns out that such virtual worlds need not even involve "coldness cues" like ice and snow; just that they allow the sufferer to travel virtually and away from their painful physical environment, strongly suggesting that the virtual reality distracted them from their pain.

Treating Post-traumatic Stress Disorder

In some circles, Albert "Skip" Rizzo is known for his rugby skills. However, as a clinical psychologist and scientist at the University of Southern California, Skip is an expert on using virtual reality to treat psychological disorders, in particular, post-traumatic stress disorder (PTSD). In May 2008, the *New Yorker* featured Rizzo's work with veterans of the Iraq War.

Rizzo uses virtual reality for "habituation" therapy; specifically, he employs the technology to repeatedly immerse someone in the scene of the original traumatic event that caused the PTSD in the first place. Habituation therapy assumes such immersion will reduce PTSD sufferer's panic when everyday events bearing a similarity to the original traumatic event occur, by preventing the everyday event from triggering painful memories. For example, if a returned soldier with PTSD were driving down an American street and saw a child running across the street somewhat ahead of him, the scene might trigger memories of a child that he saw run over by a convoy, or even one he ran over, raising anxiety, guilt, and even self-loathing. So a goal of immersion therapy is to break the connection in memory.

Rizzo developed Virtual Iraq for such habituation. He created the first version using a commercial video-game platform. Given that, by some estimates, more than 20 percent of Iraq veterans suffer from PTSD, the navy was so impressed by Rizzo's prototype that they asked him to develop a more sophisticated version. Rizzo explains Virtual Iraq is successful because young soldiers are nearly inseparable from their video games when they aren't soldiering. Consequently, the notion of using the medium with which they are most comfortable seemed intuitive.

Virtual Iraq provides therapists with many options to re-create the ambient conditions in critical trauma scenes for soldiers suffering from PTSD. They can, for example, turn day into night, whip up a sandstorm, crack a windshield, populate a market square, or increase the volume of mortar fire that's heard. Not only does Rizzo's simulation include sight and sound, it has smells, too. A virtual "olfactory" device tracks a person's movements and reproduces appropriate smells in the vicinity of the person's nose based on where the person is looking and what he is near; for example, the smell of burning hair or rotting flesh when someone approaches a virtual corpse. Rizzo's system has produced incredible treatment results. Clinical psychologists at a number of institutions are using Rizzo's treatment tool, and the success rate (measured by whether these patients still meet the clinical criteria for PTSD) has been outstanding. Many patients show marked improvement after five to six weeks of treatment.

A soldier receives PTSD treatment in Virtual Iraq.

MILITARY

The military utilized nondigital virtual reality long before there were computers. The notion of war games as training simulations is as old as the hills on which they are often enacted. Indeed, one of the best-known military historical documents is Sun Tzu's *Art of War*, which describes strategy, tactics, and simulation. Practice makes perfect, and there is perhaps no more consequential human activity that requires practice than war. Some form of war game has been used by just about every military organization in history. It is no surprise that virtual-reality technology, especially in its early years of development, has been closely tied to military projects.

This connection is not lost on Hollywood. For example, the hit 1983 film *War Games* likened the Cold War to a video game based on futility, namely Tic-Tac-Toe. The concept of fighting by virtual proxy has received attention more recently in James Cameron's blockbuster *Avatar*. Novelists have also latched on to the idea of virtual reality and war games. A generation of science-fiction fans has grown up reading Orson Scott Card's cult classic *Ender's Game*, a novel that features child prodigies using virtual-reality games to control battalions of avatars that are actually fighting light-years away.

Flight Simulators

Perhaps the quintessential example of modern virtual reality in the military is the flight simulator. Flying supersonic planes is dangerous and expensive; consequently, it is much better to weed out early-learning errors in a virtual simulation than at thirty thousand feet in the cockpit of a multimillion-dollar plane. The first well-known flight simulator was used in 1929. Designed to train pilots,

The Link Trainer, Edwin Link's flight simulator.

the Link Trainer, named after Edwin Link, was built in Binghamton, New York.

The Link simulated motion using pneumatic bellows that could rotate the mechanical cockpit in the three dimensions (vertical, horizontal, forward/backward), in which actual planes move. The simulators were successful in training pilots, and because a number of training accidents in actual planes occurred, thousands were used to train U.S. and allied pilots in the early 1940s.

Today, flight simulators have become commonplace, and, for the most part, they are digital, though, of course, the military still uses mechanical devices to simulate movement. The immersion created by flight simulators is extremely high, so much so that many FAA requirements for pilot training can be fulfilled with approved flight simulators.

Just like surgeons who can "see" hidden organs, pilots using virtual reality can "see" what is outside the plane. Head-mounted displays, in essence, make the plane transparent—by moving her head around, the pilot can see anywhere outside the plane. In essence, virtual reality can turn any vehicle into Wonder Woman's invisible plane. Moreover, the system can render hidden objects onto the pilot's field of view—for example, enemy planes picked up by radar and mountains cloaked by darkness or fog. Perhaps the best example is the Super Cockpit developed by Thomas Furness, a scientist who worked for the Air Force in the mid-1980s. His system involved a head-mounted display that feeds the pilot 3-D images of the surrounding environment without the need for a glass windshield.

Military Recruitment

Males in their late teens and early twenties love video games. As a mechanism to educate the populace about life on the battlefield, the U.S. Army has spent more than 10 million dollars to create *America's Army*, releasing the game on the first Fourth of July after 9/11. The investment paid off. *America's Army* quickly became and has remained one of the top ten most popular "shooter" games ever created.

One difference between *America's Army* and typical shooter games are the high levels of chronological and social realism. Players endure boot camp, become part of a team, earn medals, and "train" and "fight" with literally thousands of other players simultaneously via the Internet. Players must pass training requirements before qualifying for "missions," and the quality and excitement of a mission is conditional upon their previous performance. Furthermore, *America's Army* can be downloaded for free, which makes it appealing for obvious reasons.

The game has been updated numerous times and can be used on

A scene from the video game America's Army.

a number of game platforms. It has had a dramatic effect on military recruitment efforts, successful beyond the imagination of those who conceived it. The game implements two major recruitment strategies. First, it generates buzz and excitement about joining the army, and, in general, works as a positive public-relations campaign for the military. Second, game data provide important information to the army in the form of identifying promising potential recruits. Of course, the game has been extremely controversial and protests have been rampant. For example, in 2008, a large group of veterans, parents of soldiers, and concerned citizens in general protested the game-maker's headquarters, carrying slogans such as WAR IS NOT A GAME. Because the game is rated allowable for children thirteen and older, it is particularly controversial, as governments are not supposed to actively recruit anyone under the age of seventeen, according to international law.

Using the virtual-tracking techniques we mentioned when discussing digital footprints (chapter 10), it is possible to use an individual player's performance to determine what type of soldier he or she should be, as well as how good a soldier the player might be in the field. One of the game's creators, Mike Zyda, who now teaches classes on game design at the University of Southern California, describes how aptitude metrics were an important priority in the game's design. The idea is that the army would have the game records to analyze and use as a guide for vetting and attracting potential recruits.

The U.S. military is not the only one to use video games to recruit military personnel. Hezbollah, an organization considered by many in the international community to be terrorist, in 2003 released the first iteration of *Special Force*, a 3-D video game designed specifically to recruit Arabic-speaking players. Players "train" at a virtual Lebanese war college and can "wage attack" on specific figures in the Israeli government—for example, the prime minister. On the cover of the box, the designers advertise that the game is intended to simulate "the defeat of the Israeli enemy and the heroic actions taken by the heroes of the Islamic Resistance in Lebanon." On the other hand, in an interview in the *New York Times*, the game's designers point out that *Special Force* is actually much less bloody than the typical first-person shooter game.

Cultural Boot Camp

A relatively new and interesting application of virtual reality involves training soldiers on the detailed nuances of foreign cultures, where they may be deployed. When occupying foreign territories, proper adherence to cultural norms can provide substantial strategic advantages. Even gestures as simple as making eye contact with the

Cultural boot camp: A soldier prepares to interact with villagers.

right person and maintaining appropriate personal space can make or break a negotiation. For example, think about the interaction occurring between the soldier and the villager in the simulation above, produced by USC's Institute of Creative Technologies.

In it, a soldier can role-play using a variety of strategies: for example, eye contact, interpersonal distances, tones of voice, and topics of conversation. The virtual tracking equipment automatically measures all of the soldier's verbal and nonverbal actions and then interactively animates the villager to respond to those actions. By interacting with "members" of other cultures, soldiers learn implicit social rules in those cultures, and consequently can effectively implement a wide range of negotiation strategies, ones that can save lives and property.

Virtual Baghdad

Another important military application is the much-heralded Department of Defense virtual world *Urban Resolve.* Costing just over 20 million dollars to implement in 2006, this virtual training program prepares military personnel to cope with the pitfalls of urban warfare, such as improvised explosive devices, loss of communication signals, and challenges like dust, sand, and smoke. The creators produced a highly accurate virtual model of Baghdad, not just the 3-D physical structures, which comprise more than 2 million digital objects, but also the city's sociological, political, and economic forces.

The actual training event involved approximately a thousand trainees and soldiers in nineteen separate physical locations, many participating up to eight hours per day over a period of a few months. The purpose of this grand virtual event was not only to train soon-to-be-deployed soldiers but also to isolate shortfalls in technology and planning readiness. Some of the soldiers were embodied as avatars; others monitored the situation from various control centers.

It is difficult to predict the effect of an action on an entity as complex as an entire city. Using the virtual simulation, military strategists are answering questions such as "What happens when electricity is knocked out for twenty-four hours?" or "What happens to a city infrastructure when an attack accidentally knocks out a sewer plant?" Critically, virtual-reality technology simulates these complex "macro" processes, ones involving many actors with different goals.

Additionally, military personnel can gain valuable experience playing a game of urban "hide and seek." Given that insurgents sometimes hide in plain sight, it is critical to develop strategies for searching and locating the bad guys, separating them from ordinary citizens in an urban environment. By actually assigning team members various roles in the simulation, the military organizations are learning optimal techniques to carry out missions that clearly would not be possible to implement safely with physical people in an actual city.

PHOBIA TREATMENT

People fear many things—snakes, planes, and even snakes on planes. Sometimes these fears are rational. Cobras, after all, are extremely dangerous. On the other hand, sometimes such fears are irrational; for example, when one will not walk within a city block of a pet store that sells snakes. When fears debilitate people's lives, psychologists label them *phobias*. There are many types—fear of heights, public speaking, the outdoors, germs, flying, etc. One of the more heralded strategies for treating phobias is a slow and costly process called "systematic desensitization."

While we were in graduate school at the Department of Psychology at Northwestern University, we shared an office wall with a clinical laboratory that studied the systematic desensitization of those suffering from arachnophobia—fear of spiders. The lab, which produced many successful therapy techniques over the years, housed glass tanks with dozens of spiders, ranging from smaller household varieties to large, hairy tarantulas.

Ideally, during the first therapy session, the arachnophobe would enter the lab and would simply stay in the same room as a cloth-covered spider tank. During the next session, the client would again simply stay in the same room, but now the clinician removed the cloth so the person could see the spider inside. During the next session, the patient would turn toward and look at the glass tank, and so on, until, in incremental steps, the patient allowed a hairy spider to crawl up his arm. By gradually demonstrating the irrationality of the phobia via exposure to the anxiety-producing stimuli, the clinician was usually able to cure the patient.

However, over the years, a number of problems arose. Once, during a holiday weekend, the power went out because of a Chicago blizzard. All the students and professors were away. When they returned, the scramble to get the power back on in order to save the spiders (who required very specific levels of lighting and heat) was amazing to see. These efforts were only moderately successful. Not surprisingly, on a later occasion, an uproar occurred when a few of the critters escaped, demonstrating that fear of spiders is a phobia shared by more than a few professors. It goes without saying that feeding, cleaning after, and otherwise caring for the spiders can be an arduous endeavor. However, none of these problems would have occurred if the spiders were virtual.

Virtual reality has been used with outstanding effectiveness to treat phobias. The advantages of virtual over physical desensitization treatments are many. The cost and effort of maintaining a virtual spider, a virtual airplane that takes off and lands, a virtual audience of twenty people, etc., are much lower than the costs and efforts of doing so physically. Additionally, the degree of control that virtual reality provides a clinician is very precise. When desensitizing a patient who suffers from a fear of flying via virtual reality, a therapist can regulate the amount of simulated flight turbulence down to the minutest levels in terms of force and duration. Obviously, this kind of control can produce regimens of treatment not possible with live actors or most mechanical devices. Another advantage is the convenient portability of virtual reality. Such environments can be ported to literally every therapist on the planet who wants to acquire them, and can be used on patient after patient.

Research on many phobias has demonstrated the advantages of virtual reality–based desensitization therapy. Such exposure actually reduces phobias substantially more quickly than other techniques. But perhaps the most important advantage of the virtual environment for phobia treatment is safety. The therapist can ensure no harm comes to either the patient or the bystanders.

One of the most successful virtual phobia treatment centers is the Virtual Reality Medical Center, run by scientist-practitioners Brenda and Mark Wiederhold, who have established offices in California and around the world. Their center has been successfully treating a full range of phobia patients for more than a decade. Their typical patient comes in for about ten sessions.

Fear of public speaking is one of the most common social phobias. Jerry Seinfeld once joked that, as fear of public speaking is known to rank higher than fear of death, most people would rather be in the casket than giving the eulogy at a funeral. Stage fright has likely affected all of us at one time or another. A number of scientists have explored the fear of public speaking using virtual reality. The therapist can manipulate the attentiveness and behaviors of the virtual audience, which can range from amicable (e.g., nodding and paying attention to the speaker) to unsympathetic (e.g., checking their watches or falling asleep). Mel Slater, one of the pioneers of measuring social responses in virtual worlds, ran a study examining fear of public speaking. To do so, he created a virtual audience of agents that displayed "ambiguous" gestures, ones that did not clearly indicate whether they were amicable or unsympathetic.

Slater screened responses to an advertisement he placed in newspapers in order to find people who were particularly terrified of public speaking and those who were not. When the former showed up at his lab, their worst fear came true—they were instructed to speak extemporaneously to the virtual audience. Slater's speakers found themselves at the head of a conference table, facing a roomful of virtual embodied agents exhibiting very convincing nonverbal gestures, postures, etc., designed to heighten the fear of the participants. Participants spoke for five minutes on a topic they had chosen from a list—for example, the state of British politics.

Slater and his colleagues compared phobics to the confident speakers. Both types of participants spoke either to the group or to an empty virtual room. The public-speaking phobics showed more anxiety—as measured by heartbeats per minute as well as by post-experimental interviews—with the audience than in the empty room. Non-phobic speakers, on the other hand, showed no such differences.

As a consequence, Slater and colleagues concluded that virtual agents could be used effectively to treat phobias. Indeed, the Virtual Reality Medical Center treats many patients using these types of simulations, which are real enough to cause phobics to sweat, to have an increased heartbeat, and to describe themselves as being terrified, even though they know the audience is only made up of computer agents. But after receiving systematic exposure to different types of virtual audiences, public-speaking phobics can be helped.

VIRTUAL VACATIONS

Philip K. Dick's science fiction foreshadowed many of the projects that take place today in research laboratories around the world. Indeed, the line between science fiction and scientific research is often fuzzy, with scientists and engineers drawing inspiration from their favorite novels, and writers visiting science labs in order to gather ideas. Consider the two "bibles" of cyberpunk (the genre of virtual-reality fiction): *Neuromancer* and *Snow Crash*. These texts are widely read by virtual-reality researchers, and, in fact, Jaron Lanier, one of the pioneers of virtual-reality research, collaborated often with William Gibson, the author of *Neuromancer*. Philip K. Dick had been delineating major concepts regarding virtual reality for decades, even before the cyberpunk genre appeared. In 1966, Dick published the story "We Can Remember It for You Wholesale." About twenty-five years later, Arnold Schwarzenegger starred in a movie loosely based on that story, called *Total Recall*. The premise was that virtual travel is cheaper, safer, more predictable, and ultimately more enjoyable than physical travel. The plot revolves around a virtual vacation to Mars that blurs the line between the grounded and the virtual reality.

As pointed out by Jaron Lanier, one of the best outcomes of virtual reality will be saving lives that are lost in preventable traffic accidents during travel—for example, on the way to the beach or visiting relatives. The number of deaths in the United States caused by traffic accidents is staggering. Compare the annual death toll from car travel—more than forty thousand per year—to those lost in the war in Iraq. Moreover, in this era of going green, think about the savings in carbon footprint that we will achieve when jet fuel–guzzling planes and SUVs become a novelty.

Not surprisingly, scientists for the past few decades have pursued virtual tourism in a number of ways, not only for entertainment but for educational purposes as well. Donald Sanders, who received his doctorate from Columbia University in anthropology and who currently is president of the Institute for the Visualization of History, has dedicated his career to creating virtual representations of historical events and locations. For example, by analyzing texts, photographs of ruins, and countless two-dimensional reconstructions in drawings, the institute was able to produce a 3-D model of the Acropolis in Athens, Greece.

The critical issue in reconstructing historical sites is to build accurate models based on the available historical information. After pouring over often conflicting maps, documents, images, and accounts, the historians built what they believe to be a near-exact

The virtual Acropolis.

representation of the Acropolis. Other researchers were able to collaborate and virtually walk around the fifth-century site, gathering evidence to answer long-asked questions regarding ancient lines of sight and spatial configurations.

In a less academic context, the same idea can be applied to sites that still exist. For example, there is a gorgeous natural formation in Alabama, called Cathedral Cavern. For safety reasons, the local government has decided to fill it in with sand before it collapses. In anticipation of the demolition, scientists have used 3-D laser scanners and photographs in order to create a virtual model of the cave, preserving the experience for future generations, who will be able to experience the cave through immersive displays housed in museums.

Virtual tours don't always have to be based on actual physical locations. For example, many trips can be taken in *Second Life* for free. There are dozens of vacation hotspots modeled inside, and the inhabitants of *Second Life* have created a wide variety of automated tours—that is, a computer program and tour guide that virtually teleports people from one location to another while describing the vistas and answering questions that cyber-tourists may have about history.

There are virtual travel agencies that sell information about these tours, and at a bookstore anyone can purchase *The Unofficial Tourists' Guide to* Second Life, published by St. Martin's Press. In it, one can find advice on the best tours available—for example, dancing in Ibiza at a rave on the beach, taking a train ride across the countryside, flying in a hot-air balloon, swimming underwater around shipwrecks and colorful tropical fish (without scuba gear!), or even sitting on the bridge of the USS *Enterprise* from the original *Star Trek* movies. These are exciting virtual travel experiences even now. Imagine what they will be like when people use a more immersive *Second Life* that features touch and smell as well as sounds and sights.

Perhaps the biggest beneficiaries of virtual tourism will be giraffes and orcas. Although zoos play a valuable function in terms of conservation of species and research, they also function to educate and entertain the public. The animal rights group People for the Ethical Treatment of Animals (PETA) has lobbied on a number of occasions for the use of virtual reality for zoos. A letter written by Kristie Phelps, assistant director for their Animals in Entertainment campaign, sums up the values of virtual zoos:

> *Humane alternatives to these animal prisons—such as virtual-reality exhibits—can provide visitors with an exciting, educational experience. Large, open areas with immersive*

3-D video projections would allow visitors to view and interact with virtual orcas and other dolphin species. Incorporating light and motion sensors, sound sensors, computer consoles, and touch screens into the design would give visitors the opportunity to interact with virtual sea life. The experience would be like watching free-roaming wild animals, whereas captive animals' unnatural and repetitive behavior patterns—which are linked to their oppressive environments—are dull to watch and provide little educational value.

Another interesting form of tourism is the notion of a virtual window, which will undoubtedly be welcomed upon its dissemination by those stuck inside corporate cubicles across the globe. The average person takes advantage of this idea on some abstract level—for example, with a soothing screen-saver on her computer, or, alternatively, a poster of a nature scene on the wall. But research has shown that even slightly ramping up the level of immersion can have positive psychological effects, which is good news to all of us who work in non-corner offices. In an experiment, Dutch professor Wijnand Ijsselstein, who also works with the electronics mogul Philips, made a simple but elegant improvement on a virtual window by incorporating head-tracking.

You can try it at home. Walk by a window, look outside, and see how the scene changes, based on how you move your head. Every time your head moves, you see new objects through the opposite side of the window and get new information about depth as well. Next, try to do the same thing to a poster. No matter where you move your head, it always looks the same. This is an obvious difference, but a good illustration of how rich and pleasant the scenery can be outside a window simply in terms of the raw amount of information it offers.

Professor Ijsselstein designed a system for a virtual window that updates the scenery based on a viewer's head movements (i.e., monitored by tracking equipment), like an ordinary window. He also created one that did not update as a result of the viewer's head movements (i.e., like a poster). He then put experimental participants in a room, and showed them the exact same "outdoor" scene using both systems. He demonstrated that the virtual window resulted in more appreciation for the scenery than the static version. By augmenting a pleasing picture with head-tracking updates, a designer can improve the mood of those who enter a room. Of course, the opposite is probably also true. A designer can terrify people by using the same technology to project a scary scene.

VIRTUAL REALITY IS AT, OR IS RIGHT AROUND, THE CORNER FOR innumerable applications. The physicians who treat us, the companies that design products, the military that protects us, the historians that educate us, and even people in the tourist business are constantly maximizing their impact more and more every day via virtual reality. As devices become cheaper, more comfortable, and disposable, these and many other areas of application will be disseminated broadly and amaze people. Virtual reality will be considered not only a high-end research instrument, or even an entertainment platform, but a tool that is used in all sorts of daily activities.

Our children have asked us what our world was like before computers. We try to explain as they frown. Generations to come will ask their parents what it was like before virtual reality. Perhaps, those parents will answer, "Well, why don't you visit your great-great-great-grandfather's avatar and find out?!"

CHAPTER THIRTEEN

VIRTUAL YIN AND YANG

The "yin and yang" dichotomy refers to opposing forces, commonly depicted by the familiar black-and-white symbol. According to ancient Chinese philosophy, yin and yang are constantly active, pushing and pulling against each other within every known system. In Western thinking, yin and yang typically refer to good and evil forces. Here, they refer to the contrasting moral consequences—pro-social versus antisocial—of virtual reality. We explore the manners in which virtual reality has and will affect societal institutions. Yin-and-yang issues are better known to scientists as the "dual-use problem" referring to the argument that any technology can be used for good or for evil, certainly including virtual reality.

DUAL USE

It's not quite clear when "manufactured" physical tools were first devised on Earth, but archaeologists estimate that they emerged at least

2 million years ago, during the Stone Age, among the great apes (i.e., hominids), the predecessors of whom we recognize today as chimpanzees, gorillas, orangutans, and humans. Most early or primitive tools were developed primarily to secure food via hunting, gathering, and scavenging. Later, the Bronze Age and the Iron Age marked progression of tool-making materials from rock and stone to metals. Only in the very recent past, archaeologically speaking, have tools been made from other materials, such as rubber, plastics, and silicone, as well as a raft of more exotic ones, like titanium alloys and carbon fiber.

It's also murky when or even how language, primarily a psychosocial tool, originated. Many animals are capable of speech (e.g., "Polly wants a cracker!") as well as communication (birdsongs and whale sounds), and at least one chimpanzee, named Washoe, and one gorilla, named Koko, have been taught American Sign Language successfully. No one quite knows how verbal language came about, though most scholars assume that language developed somehow via evolutionary processes.

Interestingly, both physical tools and linguistic ones have nearly always qualified as "dual-use instruments." The sharply honed stone tied to the end of a stick could be used to spear a fish or mammal to provide food for the residents of a village or, on the other hand, to invade a distant enclave to steal provisions and kidnap women and children. Analogously, language could be used to bring people together for purposes of survival, charity, public works, or problem-solving, or used to keep people apart via ostracism or the threat of aggression. George Orwell's novel *1984*, in particular, highlights how a tyrannical government can use language to sway the masses, for example the famous slogans WAR IS PEACE, FREEDOM IS SLAVERY, and IGNORANCE IS STRENGTH.

Moving forward to the present, one realizes that things have not changed much in at least one important respect. Physical and linguistic tools still are dual-use instruments. For example, although nuclear fission and fusion technologies underlie doomsday nuclear weapons, they also have peaceful uses, such as the generation of household electricity and medical imaging of internal body organs. Linguistically, slants are used to euphemize or denigrate actions, e.g., the "Patriot Act" instead of "suspension of civil rights," or "socialized medicine" instead of "medical insurance reform."

Communication media technologies have "dual uses" as well. With each technological advancement, new media technologies have magnified existing social effects, some of them positive and others negative.

THE DIGITALLY GOOD, BAD, AND UGLY

Reflect on the following, relatively recent examples of the Internet's social effects. Later in the chapter, we examine how, once the technology matures into fully immersive virtual reality, these consequences will be magnified.

Joe Trippi is a card! A political guru and Internet-technology revolutionary, Trippi once regaled us at a dinner party with his funny recollections of being on the campaign trail during various primary-election campaigns.

It was Joe Trippi, a veteran campaign manager for high-office seekers in the United States and abroad, who managed Democratic candidate Howard Dean's 2004 presidential campaign. Although Dean lost the campaign, Joe Trippi won the war. In less than a year, Trippi and his cohort transformed long-shot candidate Dean's limited assets. A little more than 400 identifiable supporters and a campaign coffer of $100,000 grew to more than 640,000 supporters and a financial war chest of more than $50 million, more money than that of any primary candidate to that point in U.S. history. The Internet made Dean a formidable candidate. Most of his donations "dribbled in" through individual contributions of $100 or less. Four years later, Trippi's work for Howard Dean served as a model for the Obama campaign, which eventually had a base of 13 million people and raised a record $800 million from 5 million donors, an average of less than $200 per donor, giving millions of voters a collective chance to have a significant impact on election outcomes.

Of course, there are many other positive impacts that the Internet has had on societies across the world. People are more connected with others than ever, and it is heartening to build mountains of financial aid via small donations to others in areas devastated by earthquakes, floods, tsunamis, etc. If information is power, more and more people have more and more information literally at their fingertips than anyone imagined even two decades ago.

People have more and more reason to worry that their identity may be stolen by Internet thieves, primarily for financial gain. It is estimated that more than one in eight people in the United States have experienced such problems. How easy is identity theft? Very easy. In broad daylight or during the dark of night, the identity thief parks her car near the potential victim's abode. Within the relative safety of the automobile, the perpetrator hacks into the hapless mark's wireless network (too often an unsecured one) and easily finds the information she needs.

The victims are none the wiser until their checking account balance is overdrawn or a five-figure charge for a month in Tahiti appears on their credit card statement. It takes a victim of identity theft only a few minutes to lose crucial private financial information, and not much more time than that to have her financial reputation ruined. It takes the victim many months and sometimes even years to reestablish credit and recover lost funds if they are ever recovered at all.

In October 2006, Tina Meier, a resident of Missouri, found her thirteen-year-old daughter Megan in the latter's bedroom, where she had hanged herself. Megan died the next day. Tina and Megan's father, Ron, determined that a virtual but very real drama for Megan and her parents had played out on the Internet. Megan had been communicating via MySpace with a "boy" who had expressed interest in being her friend. Megan's parents later learned that she had received an e-mail from this "boy," "Josh Evans," telling her that he was no longer interested in someone as bad as her. Megan had also discovered that there were all sorts of hurtful Internet postings about her that were quite negative.

In due course, Tina found out there was no "Josh" and that the e-mail correspondence was concocted by Lori Drew, the mother of one of Megan's former female friends. Drew was eventually indicted and charged with conspiracy and multiple counts of accessing a computer without authorization via interstate commerce to obtain information to inflict emotional distress. Ultimately, she was acquitted. Ms. Drew admitted creating the "Josh Evans" MySpace account with her daughter and one of Drew's employees. At Lori Drew's trial, witnesses testified that the conspirators used "Josh" to get to know Megan so they could humiliate her in retaliation for Megan allegedly spreading gossip about Drew's daughter.

Could these three events have occurred even ten years ago, say at the turn of the millennium? It is doubtful. In 2000, there were about 100 million Internet users worldwide. Few politicians thought that the Internet could help their campaigns. Some financial fraud occurred online but nowhere near the level we see today. And social-networking sites were nonexistent.

But now nearly a third of the world's population is wired. Most monetary instruments, including "cash," exist only in electronic form stored on silicon. Social-networking sites are highly populated. The line "If you build it, they will come," from the baseball-themed movie *Field of Dreams*, is illustrated by Silicon Valley entrepreneurs such as Brad Greenspan, who oversaw the launch of MySpace, and Mark Zuckerberg, the cofounder and CEO of Facebook and *Time* magazine's 2010 Person of the Year.

THE FUTURE

In the past fifteen years, virtual reality has accelerated at an astounding pace—it seems that almost every month agents and avatars can be built faster, cheaper, and better. While there may not yet be a Moore's law for avatar advancements, the rate of change over the past few years leads one to project that in a few decades the norms of social interaction, war, education, sex, and relationships will change drastically. Predicting this future is risky business; just ask any scientist from the 1960s who guaranteed the skies would be plagued with flying cars and homes with robotic servants. Nevertheless, we can illustrate what the future might hold for our children and grandchildren. We do not do so blindly; instead, we will look to the professional futurists: science fiction authors. There is a long and intertwined relationship between virtual-reality researchers and the latter.

In doing so, we are forced to make assumptions about avatars. Although we may be wrong about some of these assumptions, for the purposes of this discussion, let's assume that they will come true.

First, avatars will become perceptually indistinguishable from their flesh-and-blood counterparts in terms of how they look, sound, feel, and smell. This is far from an outlandish claim, given how far virtual representations have come in the past decade. Computer graphics used in movies have established that avatars can look just like their owners. For example, in order to artificially age Brad Pitt in *The Curious Case of Benjamin Button*, the filmmakers constructed a near-perfect digital replica of his head and face. Sound, smell, feel, and even taste are sure to follow as the years pass. To put this in perspective, forty years before this book was written, there was no Internet or cellular phones, and twenty years before, there were fewer than fifty Web sites.

Second, people's control of their avatars will become automatic, without having to use any type of controller, keyboard, or even vocalization. Avatars will be manipulated via people's everyday actions or thoughts, much like in *The Matrix*, where Neo's avatar interacted with his brain via a neural jack plugged into the back of his head. A similar control was described in William Gibson's *Neuromancer*, considered by many to be the first thorough description of virtual reality. Wearing an avatar will be like wearing contact lenses, with people unaware that they are wearing anything. In the same way that Nintendo Wii has made a huge jump in avatar control over the joystick and keyboard, as the decades pass, avatars will be controlled as easily as our own physical bodies.

Third, avatars will be embodied and capable of touch. Whether via haptic devices attached to digital simulations, robots tele-operated by owners, or even biological clones that people's minds can control, avatars will "walk among us" and interact with others who are present physically rather than digitally. Today we see blending of the digital and the physical—for example, kids walking down the street like zombies (or in the words of Neal Stephenson, author of *Snow Crash*, "Gargoyles") as they text and game within virtual worlds. Forty years from now, such blending will be routine, and humans and avatars will jostle one another in shopping malls.

Let's consider the impact of virtual reality, present and future, on a sample of four well-known social institutions (close relationships, religion, education, and national defense). With some trepidation, we speculate about the possible impact of virtual reality on them in the future near and far. And let's remember: "The Future" seems to come about more quickly these days than it did years ago.

BEING TOGETHER

Among the strongest bonds within social networks are those that occur in close relationships, especially intimate and familial ones—

between married couples, parents and children, siblings, etc. There are, of course, cultural variations in familial relationships. Marriages in some cultures are monogamous (single spouses) while others are polygamous (multiple wives), and even a few are polyandrous (multiple husbands). There is an ever-strengthening social movement in the United States to institutionalize same-sex marriage and an equally strong one to prevent it—an interesting yin and yang in itself. But the possibilities that virtual reality brings to close relationships will stretch the limits.

We've shown that romantic relationships, for all their diversity, work in virtual reality in ways surprisingly similar to grounded reality. One behavior expected of romantic partners is what psychologists call "social support." When people are particularly stressed by events, they need to know that their partner will be there. However, for various reasons, romantic partners may not be accommodating every time their partners need help. This can make for less than a smooth romantic relationship.

Often, couples engage in activities to keep the romantic fires burning. Some couples take ballroom dancing lessons, others go to swingers clubs. Still others answer advertisements to participate in a virtual-reality experiment at the University of California, Santa Barbara. We studied the latter couples in an experiment on how romantic partners function in virtual reality.

Collaborating with Nancy Collins, a professor at UCSB and one of the world's top experts on the psychology of close relationships, we created and examined what would happen in virtual reality when relationship partners provided or did not provide social support. These couples received a big surprise on arrival at the lab—they were placed in one of the more stressful virtual realities we have created. We wanted to see how people reacted to their partners after receiving—or, unfortunately, not receiving—

support from them during a stressful situation. Upon the arrival of each couple, we immersed one partner in virtual reality. That partner found himself or herself standing on a harrowing trail that wound along a mountainside with sheer cliffs rising above the trail on the left and a sheer drop on the right. At the end of the trail, fifty virtual feet or so away, the hiking partner saw his or her romantic partner's avatar standing in a small clearing bounded on the left by the mountain wall and fenced in on the cliff side, complete with a picnic table—quite a romantic scene.

As investigators, we randomly manipulated whether the romantic partner's avatar socially supported the hiking partner. In the attentive condition, they socially supported their partners, by waving to the hikers and nonverbally communicating, "Come on, honey, you will be fine!" and beckoning them to walk along the terrifying and harrowing trail to reach the romantic picnic spot. In the inattentive condition, the partner simply stood there staring away over the chasm and enjoying the view, seemingly communicating, "Everyone for themselves!" In a third, control condition, the partner was not present at the romantic spot at all (i.e., no avatar).

Of course, participants knew before they entered virtual reality that there was no corresponding cliff in physical reality, but, as to almost everyone who has been through a pit demo, it didn't matter. People were obviously frightened as they walked toward the partner. (Having vultures circle and small rocks roll past and fall off the path and over the cliff as participants stepped near them was quite effective in this regard.) Participants reacted to their partner's support or lack of support in the virtual world as they would in the physical world. Not surprisingly, the data showed that the least stressed hikers were those who had attentive, supportive partners.

You can do it! A participant sees her romantic partner's avatar beckoning her to the picnic spot.

However, the key question, for our purposes, was examining what would happen to the relationships after the experiment ended. Back in grounded reality, the virtual hikers whose partners had been inattentive (recall that, as experimenters, we actually manipulated who was attentive and who was not) maintained more personal space between themselves and their partners, staying farther away from them than they normally would have. This lengthened personal distance indicates a decrease, at least temporarily, in the intimacy of the relationship. Because the partners did not provide support to their loved ones, they earned a stay in the "doghouse" upon return to grounded reality. As scientists, we were pleased that our results verified the hypothesis that virtual behaviors affected physical relationships.

When we talk with people about the possibility of intimate relationships in virtual reality, they invariably bring up the issue of touch. Perceptual scientists use the term *haptic*, which means "related to the sense of touch." Let's face it—in close-relationship situations, people would like haptics in their virtual-reality experiences. They want to touch and be touched by romantic partners.

It is not yet easy to provide direct haptic experiences in virtual reality. Nevertheless, it is not surprising that the sex and pornography industry has come up with all sorts of robotic virtual devices, so-called sex aids, available widely in the Internet marketplace, including vibrators, dildos, artificial vaginas, etc. But users still have to operate these devices themselves. Or do they? There is an emerging field called *teledildonics*. That label refers to the creation of haptic sexually stimulating devices that can be controlled by others via the Internet. Teledildonics makes it possible that "if you can't be with the one you love," you can still be with them in virtual reality, "haptically" ever after.

Of course, haptics are important in relationships for more than sexual reasons. People touch each other in nonsexual ways, too. For example, physicians touch or palpate patients' abdomens during a general physical exam. Some parents sometimes spank their children. People often touch each other while dancing. And, at least in Western societies, people shake hands! There are technologies available to support all of these kinds of touches, which can be used in virtual reality.

Haptic devices are becoming more common—in addition to the device for handshaking, described in chapter 8, we are working with Philips, the Dutch electronics company, to add a "touch layer" to telephone conversations so that intimate conversations can become, well, more intimate.

So far, all of this is to the good, although some readers may have a disturbing reaction to the emerging field of teledildonics. Virtual reality expands the possibility of close relationships with others located anywhere in the physical world. However, the dual-use principle tells us there is a downside to virtual reality for close relationships as well. We've talked about Internet addictions, including sexual ones, which steal time from individuals' lives, and that can be as problematic as infidelity. Recall the report of a *Second Life* relationship leading to the dissolution of the marriage of a British couple. But the problem gets more complicated.

One of our students had an interaction within *Second Life* that raises fascinating issues regarding monogamy. He met a woman in *Second Life* who had a husband in physical reality. She and her husband spent lots of time in *Second Life*; indeed, they first met there before meeting in the physical world and subsequently getting married. When they were together in *Second Life*, she and her husband always wore the same avatars. On the day she met our student, she was wearing the avatar that she typically uses when interacting with her husband, though he wasn't with her. She and our student talked, flirted, and somewhere along the way, one of the two raised the prospect of engaging in virtual sexual activity (similar to phone sex, but with avatars who can perform various actions with the touch of a button in addition to the voice communication). The woman was clearly interested, as she and her husband had previously agreed to a "sexually open marriage" when they were online. But they had a rule about online sex with other partners. They could only do so if they "switched avatars." In other words, she and her husband were allowed to have sex with other people, virtually, but not using the avatars they used online together. According to their rule, extramarital sex was not "cheating" if the spouse used a different avatar.

NOT SURPRISINGLY, SCIENCE FICTION WRITERS HAVE EXAMINED how sex might change in the virtual worlds of the future, even more than it has changed already. Some of the revelations are quite positive, others downright unseemly. At any rate, the constructs of sexual relations in a world inhabited by avatars—ones that are perceptually indistinguishable from actual people, controlled automatically, capable of touch, and disposable, as well—is not one for the meek.

In Gibson's *Neuromancer*, one of the most disturbing constructs is the idea of a "meat puppet." If one can build a perfect perceptual replica of another human being—either confined to digital space or alternatively downloaded into a waiting biological body—people risk losing control over their own representations. In *Neuromancer*, this concept takes many forms. For example, Molly, the main female character, experiences tough times trying to earn money. She employs a form of virtual prostitution by temporarily wiping out her own control and awareness, replacing them with a virtual program that takes over her bodily functions. Molly became a prostitute by renting out her body while her mind was dormant. A john would select her from a lineup, insert his own "behavior control" into her body, and have his way with her. After the session was over, Molly would snap back to consciousness unaware of whatever transgressions took place, good or bad.

As another example, the main male character of *Neuromancer*, Case, received the "hard sell" from someone trying to persuade him to agree to a proposal. Case, typically strong of will, nearly succumbs to a particular temptation. While jacked into virtual reality, he runs into the love of his life, who had recently passed away. Someone had reconstructed her exact body and personality. Case literally cannot tell if she is real or not; and when she offers him love and comfort, he falls for her once again. When, after some time has passed, he hears the persuasive proposal not from a third party but from his one true love, who unwillingly has pervaded his consciousness, he is almost powerless to resist.

Think about a world in which imposters can perfectly re-create and control other people's avatars. This is not so far-fetched, if you recall that Orville Redenbacher's family brought him back from the dead to work as an actor in advertisements. Imagine what will happen when anyone, at any time, can be made to perform any action fathomable. The notion of one's own body will take on new meaning, and, sadly, the slogan "No means no!" will need to be revisited. Avatars will not only be available to imposters, but they will also be fluid and transformable. Imagine the dismay of people who, like the lead character, Takeshi Kovacs, in Richard Morgan's novel *Altered Carbon*, find out that their partners actually look nothing like their avatars.

FROM THE PROFANE TO THE SACRED

> *New technologies and the progress they bring can make it impossible to distinguish truth from illusion and can lead to a confusion between reality and virtual reality . . . and risk indifference toward real life.*
>
> —Pope Benedict XVI

Early in human history, storytelling unified people institutionally. Mythical stories allowed like-minded people to share an explanation of Creation. As time passed, new communication media technologies, such as painting and sculpture, manuscripts and printed books, and eventually radio and television, allowed people to pass the myths on to succeeding generations, although often in somewhat revised forms, thereby continually reproducing religious institutions long after the founders passed away. It is difficult to think of any communication media technology that hasn't been used to help people maintain religious beliefs and to proselytize nonbelievers.

One wonders what Saint Paul, who provided religious guidance to early Christians via his epistles (i.e., letters), would have done with e-mail; how the Bible, Koran, and other ancient religious works would have become "best-sellers" without manuscripts and the printing press; and what funds would have been raised by televangelists without television. Most God-fearing people don't seem to be up in arms about any of those communicative media developments being used for religious purposes.

Both the Internet and virtual reality have even been applied to religious institutions. A recent Google search of "virtual church," "virtual synagogue," and "virtual mosque" turned up more than 24,500, 6,000, and 7,000 sites, respectively, including both 2- and 3-D digital facilities and congregations!

However—and surprisingly, perhaps, for institutions based mostly on a conception of life on earth as being temporary, or virtual, compared to the absolute reality that awaits people after death—there is serious debate about the religious value of "attending church" via virtual reality. Pastor Bob Hyatt of the Evergreen Community Church in Portland, Oregon, argues: "The worship, equipping, and discipling ministries of the church simply can't take place through the Internet. . . ." Douglas Estes, a pastor of a brick-and-mortar church in San Jose, California, and the author of the book *SimChurch*, counters: "A myth is growing . . . that online church is not good . . . Isn't church supposed to be about people in communion with God rather than the building?"

In any event, certainly, much good has come from any of the religious traditions; for example, moral codes and charity to those who need it. However, religious beliefs also have given rise to plenty of destructive forces. Many historians have traced the role of religious beliefs in causing great harm, including the torture of heretics, the subjugation of women, and all-out wars. The record of religion-based calamities is as long as human history itself.

As virtual reality becomes more immersive, both beneficial and destructive aspects of religion will be magnified. How does recruitment of a jihadist suicide-bomber change when he can put on a head-mounted display and really get a virtual taste of an afterlife with seventy-two virgins, for example? How will charitable donations change when church members can actually smell and taste the poverty that a small, third-world village is riddled with? In elementary school, we got our first taste of the Bible through a "Picture Bible," a wonderfully crafted cartoon version of the major books. It helped us to visualize the historical events that were so critical to its messages. On the other hand, certain events remain hopelessly abstract. Children in the future will have an entirely different conceptual experience of, for example, Moses parting the Red Sea. They will feel the cool water temperature, smell the fish flopping on the trail, and feel the rumblings of the waves held in check.

The notion of an afterlife is the cornerstone of most religions. There is a yin and yang to it for nearly all of them—for example, heaven and hell in Christianity. Typically, these constructs remain abstract and fuzzy for believers. Given that religious institutions have utilized every prior form of media, virtual reality seems likely to be used to create experiential afterlives—dare we say like "heaven" or "hell" on earth. Although Dante produced descriptive prose depicting the nature of hell in *Inferno*, virtual reality can enable his modern-day disciples to sample—that is, see, hear, touch, taste, and smell—his nine "circles" of hell. For one reason or another, such as atoning for sins, Christians may end up spending time in the place they are ultimately trying to avoid.

But if one can design a world that is so perfect that it can be considered heaven, then the ultimate goal of many religions—to end up in the preferred location after death—may be changed to spending time there while alive. Instead of being a good person to gain entrance to heaven at a later date, people will want to be a good person to spend time there now.

In the Old Testament, God punished and rewarded people during their lives. The Book of Deuteronomy is full of specific examples of what happens to those who engage in various behaviors deemed to be unholy. In the New Testament, the reward/punishment paradigm shifted so that the impact of one's actions was not felt until the afterlife. As avatars become commonplace, the notion of afterlife may shift once more back to current life, so that one (virtually) experiences the ultimate—heaven and hell—while still alive. This poses difficulties for religious beliefs of an intangible, abstract afterlife.

GET AN EDUCATION!

Our fathers and grandfathers, like so many of their generation, liked to talk about the good old days when "men were men and boys and girls were men," too. They emphatically explained how they didn't have cars to get to school, nor were there any buses. Teachers were strict and allowed to use corporal punishment. God help you if you didn't do your homework. Apparently, people were tougher back then, and kids today have it way too easy.

Contrast our forebears' educational environments with future ones envisioned by science fiction author Neil Stephenson in *The Diamond Age*. The protagonist, a working-class young girl, Nell, is the lucky recipient of *A Young Lady's Illustrated Primer*, which, as it turns out, is a virtual tutoring system. The system integrates virtual-reality, nanotech, and artificial-intelligence technologies to provide the most thorough education a student could ever hope for. Talk about new generations having it easy!

Will Nell's primer stay science fiction? We strongly doubt it. History argues otherwise. This transition from physical- to digital-based learning environments is becoming more prevalent for other educational objectives, and is here to stay, at least until scientists, like the ones depicted in the Wachowski Brothers' *The Matrix* and James Cameron's *Avatar*, figure out how to program brains directly. Virtual environments can add many learning tools to education.

Imagine you are a fifth-grader, reading a textbook about the Civil War—not necessarily the most motivating way to learn history. A teacher might decide that having her students watch a movie depicting colonial times might be a bit more engaging. Still more immersive might be a field trip to historic colonial Williamsburg, Virginia, where students would be surrounded by actors who can actually interact with them in a historical context.

However, in virtual reality, students can experience a level of engagement in a colonial Williamsburg reenactment without the physical travel. Moreover, there would be no crowds of educational field-trippers getting in the students' way. In fact, every sixth-grade student in the entire country could be immersed in the simulation simultaneously and feel like he had the entire place to himself!

Perhaps the most transformative feature is the capability for students to be safely present in educationally important situations. Students can be immersed in virtual environments that can be perceived as dangerous but without the actual physical danger. Once these lessons are produced, they immediately become cost-effective, as the same virtual worlds can be used over and over and over again.

For example, in virtual reality, American history students might find themselves in the middle of a Civil War battlefield, experiencing bloody combat firsthand. Astronomy students could jump to the surface of one of Jupiter's moons and better appreciate the scale and remoteness of our solar system. Political science students could witness the signing of the Magna Carta or experience travel on a slave ship and living in slave quarters on historically accurate Southern plantations.

Chris Dede at Harvard University is one of the leading researchers investigating the viability of using virtual reality for learning. Dede has been creating virtual-reality learning scenarios for more than a decade. In his 2009 *Science* article, Dede discussed the educational advantages of immersing a student "within" a lesson housed in a virtual world.

For example, Dede's studies demonstrate that students who face a medical science dilemma, such as an outbreak of illness in his virtual *River City*, learn more about epidemiology and prevention than they do in traditional classroom settings. One of the reasons for this happy result is that the students find the virtual experience more engaging. In particular, the immersion experience causes learners to spend more time "in" the lesson and thereby "with" the information contained in it. Dede argues that students reach a fuller understanding of the relationships among causes and effects via firsthand experience when they are placed in a virtual world.

Within virtual reality, students can work much more collaboratively. In the *River City*'s epidemic, students had to figure out why people got sick and what actions might contain the spread of the disease. Immersed in virtual reality, students went out and spoke to ill townspeople, hospital staff, merchants, and university scientists. By sharing their knowledge, they were able to uncover the mechanisms behind the outbreak.

Many studies have compared learning in virtual worlds like *River City* to traditional classroom settings, board games, etc. Immersion in virtual reality appears to be the key to enhanced learning effects. This is especially the case for "low-performing" students who have a hard time being engaged in traditional course materials.

Unlike traditional classroom settings, no child need ever slip through the cracks in virtual reality. One of the more difficult and controversial aspects of education, one that intensified under the "No Child Left Behind" act supported by both President George W. Bush and the late Senator Ted Kennedy, is the requirement for standardized student evaluation. The issue of standardized tests is polarizing, with proponents arguing for uniform tests and detractors claiming such tests are inaccurate and have little educational value.

Given that every student action, utterance, and gesture can be tracked and recorded in virtual reality, measures of student evaluations based on the tracking data can be stunningly precise and, consequently, more valid. In college courses at large universities, students are typically tested twice: a midterm and final examinations. Even with more frequent testing, only a few evaluative data points are possible. However, the number of informative data points available in virtual courses is many orders of magnitude higher.

Let's explain. In virtual reality, students' behaviors provide verbal and nonverbal clues to their comprehension and performance dozens of times per second. More important, this information can identify who understands the lesson and who does not. In essence, the entire course comes with a built-in continuous evaluation opportunity for those in charge of making sure that no child is left behind. Instead of just reacting to a poor test score, the system can check whether a student is gazing at the wall during a lecture, or displaying a confused facial expression. Consequently, virtual reality is more constantly attentive and interactive than a human teacher.

Given what we know about and have described previously as "transformed social interactions" and their impact on learning, we look forward to a future in which everyone can be one of Stephenson's "Nells," each with his or her own personalized virtual-reality tutoring system. In the future, students will be able to wear inexpensive headsets plugged into small devices, much like today's iPhones, which they will be able to use for telecommunications, research queries, entertainment, etc. The diffusion of knowledge will be tailored to students' ideal learning styles.

We suspect we will tell our grandchildren and great-grandchildren stories about how, when we went to school, all we had were calculators, word processors, and fallible human teachers; how school was tougher back then than it is for them. They will have the entirety of human knowledge and arts literally available in the palm of their hand.

OF ALL THE LINES IN *THE MATRIX*, NONE IS MORE SALIENT HERE than Neo's tentative declaration, "I know kung fu?" With a simple download, Neo became a master of martial arts and participated in death-defying combat scenes. Can information magically get injected into a brain? Perhaps not, but there are many reasons to believe that education will be a fundamentally different experience in the future via virtual reality.

Currently, the path to a great education involves hard work, dedication, and, of course, the mental raw materials to absorb all knowledge. In a world where an avatar can operate while its owner is sleeping, those who excel in the classroom may be characterized more by their status and management skills than by their intelligence quotients. If "learning" is simply a matter of having one's avatar repeat actions or, even more simply, a matter of injecting knowledge through computer software, then those with the resources to purchase software and the ability to keep their avatars active 24/7 are the ones who will reap the most rewards through "learning."

Embodied cognition is the psychological perspective that knowledge is stored not only in the brain but is prompted by peripheral bodily actions such as postures, gestures, and expressions. For example, Michael Spivey, while a professor at Cornell University, determined that people who solved a particularly difficult brain-teaser exhibited a set pattern of eye movements—ones that corresponded to the movements in gaze required to solve the problem. In other words, mentally, people who moved their eyes in a specific way tended to solve the problem better. A follow-up study by Alejandro Lleras at the University of Illinois demonstrated that people trained to do the same eye movements performed better on the brain-teaser than those who were not. The movements focused learners' attention more efficiently. Even though the problem-solvers thought the eye-movement task was distracting and precluded success on the brain-teaser, the opposite was true. Matching eye movements of novice learners to proven successful learners caused novices to learn faster. Others have determined that having people move facial muscles related to smiling or frowning without knowledge that they were doing so produced more or less liking for objects they were viewing, respectively.

The concept of embodied cognition is fascinating, especially given the assumption that virtual-reality technology can directly interface with brain and body. Software could be written to purposely control avatars' peripheral bodily movements during any learning task. Moreover, learners will be able to feel those movements as if they actually performed them. If repetition of movements is crucial, then learning could be improved automatically and unconsciously. Learning could take place while sitting on a couch, watching television, or even during naps, because the machine is controlling one's motor movements!

SECURITY

The root causes of war vary from religious and moral beliefs to economic gain and loss. Most wars have been fought over and on geographic territory, not to mention the resources available within them. Wars have been fought between nation-states, alliances among nation-states, and even empires. Wars have also been fought within nation-states via revolutions and civil wars.

Cyberspace makes the idea of national borders obsolete. In *Neuromancer*, William Gibson describes how virtual reality makes the idea of a border laughable. Today, nation-independent groups ranging from legitimate corporate structures to drug cartels to terrorist groups are networked via the Internet. In some groups, many, sometimes even most, members have never been in physical contact. Rather, members have coordinated their operations, accessing resources and delivering products ranging from heroin to death and destruction, via the most modern and up-to-date digital communication technologies.

In his book *An Army of Davids: How Markets and Technology Empower Ordinary People to Beat Big Media, Big Government, and Other Goliaths*, Glenn Reynolds spells out some of the details about this trend. Joe Trippi anticipated Reynolds's notions in 2004, when he launched Democratic primary candidate Howard Dean's e-mail fund-raising campaign. Joe postulates what Biblical David's slingshot would look like today if he were after a modern Goliath. A modern David's "slingshot" would allow him to contact hundreds of millions of other "Davids" around the world in a matter of seconds, track where the other Davids are and what they are doing, watch video news bulletins, disperse or receive instructions for action along with accompanying maps and photographs, and summon destructive forces stronger than opposing defenses in an instant. Trippi was talking, of course, about modern "smart phones," devices that become more capable weekly!

In a very real sense, the right to own smart phones may become recognized as a human right. In unlikely places—Iran and China, for example—smart phones help provide citizens with rights to assembly and a free press. Even in the United States, rights to smart phones can be tied to the First and Second Amendments of the Bill of Rights. However, like all technologies, cell phones are subject to the dual-use problem. For as many lost, stranded, or injured people who are rescued or helped because of cell phones, there are people who are hurt, ostracized, or embezzled because of them. Indeed, the 9/11 attackers coordinated their terrorist acts on the doomed airplanes via cell phones.

Arguably, today the byte is mightier than the pen and sword combined. The control, use, and monitoring of those practically infinite bytes is, in many ways, the current generation's battlefield, virtual though it is. Because of the Internet, communication is no longer a problem even for small and geographically dispersed networks, such as Al Qaeda. Political scientists point to the Zapatista rebellion in Mexico and the global effort to rid the world of land mines as two examples of how digital technology has facilitated social upheaval. Online, people who would otherwise feel marginalized can relatively easily find each other and coalesce into groups that take action. In the United States, the so-called Tea Party movement and Moveon .org are good examples.

A decade ago, some sociologists and political scientists argued that social movements without face-to-face interaction are unstable and likely to die. Though history appears to prove their thesis wrong, virtual reality allows for the face-to-face interaction to which these critics refer. It won't be long before a video camera in a smart phone can track a person's nonverbal gestures and facial expressions and display them via video or even lightweight head-mounted displays to one another in one-on-one and group meetings. In essence, virtual reality will combine all of the connectivity of a smart phone with all of the intimacy and power of a face-to-face interaction. People will literally be able to be in dozens of places simultaneously.

Does networked virtual reality spell the end of civilization as we know it? Will Armageddon arrive via the Internet? No one really knows. But, as we have written, dual use of any technology works both ways, good to bad and bad to good. Just as there are good guys and bad guys in physical reality, they exist in virtual reality. The agencies tasked with protecting the people of many nations, as we found out as consultants to a few security organizations, are constantly working on more ways to identify, infiltrate, and stop radical groups, including terrorists who are networked virtually.

As discussed earlier, "digital footprints" can help identify individuals, thereby providing an important weapon against crime and terror plots. When people use any digital device, even a cell phone or an e-mail program, they leave clues about their identity and intentions. As virtual reality develops further, the use of the tracking data to locate dissidents will continue to be a hot topic, both because of the promise of using it to genuinely catch bad guys and the clear threat it poses to the privacy of the innocent.

THE NATURE OF VIOLENCE IS CONSTRAINED IN MODERN SOCIETY BY a simple fact—when a soldier's body dies, he dies. In virtual reality, this is decidedly untrue. Once an avatar is built, it is trivially cheap to make digital copies, and not unforeseeable to make robotic and biological clones in the future.

On the positive side, wars could be fought without physical human casualty. Just as drones fly around during conflicts today, the soldier herself via an avatar will be operated remotely. This is seemingly great news in terms of wars being fought potentially without loss of human life and limb.

But, one wonders, "What is the point?" When violence is coupled with the notion that there is no potential harm to human life, what happens? How do humans behave when they are essentially bulletproof twenty-four hours a day? The world crafted by Richard Morgan in *Altered Carbon* takes this idea to frightening levels. Present-day war atrocities are commonplace. Even in the United States, some levels of discomfort are permitted during interrogation. The point at which discomfort becomes "torture" is the focal point of much debate. However, there is little if any doubt that actual tissue damage constitutes torture. If one's consciousness can hop from one body to the next, then others can do us harm without any worry of doing any permanent damage. In an extremely powerful and disturbing scene, Takeshi Kovacs, the protagonist of *Altered Carbon*, is tortured repeatedly as his transgressors destroy one biologically cloned body after another while he is forced to be conscious within each one. Similarly, in a world in which suicide bombing doesn't result in actual physical death, does suicide bombing become less despised?

There is evidence that even during tele-operation one can damage oneself. Currently, drone operators for the U.S. military are susceptible to post-traumatic stress disorder. Simply by operating a drone that kills enemies, a soldier can sustain substantial psychological damage. Similarly, it remains to be seen how harm caused to one's avatar affects the person behind it. As Morpheus states in *The Matrix*, when asked what happens when one's avatar is killed, "The body cannot live without the mind."

THE FUTURE OF SOCIAL INSTITUTIONS

Virtual reality is and will keep impacting social institutions, including ones other than those we have discussed above. Each year, millions of new users connect with one another via the Internet. As the number of dyadic, small-group, and very large-group ties and networks continues to expand, the world becomes a subjectively smaller place. But virtual reality won't just bring people together; it will also change the very nature of the activities and institutions that motivate all social interaction. Avatars are a game-changer in every human domain, including but not limited to love, war, worship, and learning. These social institutions have been around as long as humans have been sentient, and will survive.

Or will they? We have taken a shot at predicting some human and social impacts of virtual reality. Whether the specific prognostications we made will occur or not is somewhat beside the point. But there is a general prediction we can make that is quite sobering. Barring worldwide catastrophes such as nuclear conflagration, collision of Earth with a large asteroid, or runaway global warming, sooner or later technology will enable anybody to interact with anybody else in the world, positively or negatively, via virtual reality. People will have many tools available to influence one another in ways impossible in the physical world. And, just for good measure, everything everybody does will be archived, potentially making privacy an archaic concept.

CHAPTER FOURTEEN

MORE HUMAN THAN HUMAN

THE VIRTUAL REVOLUTION WILL LIKELY SHATTER MANY ASSUMPtions about human nature itself. Lewis Carroll's fantasies in *Alice's Adventures in Wonderland* and *Through the Looking Glass* will seem tame. In this concluding chapter, we examine the big questions virtual reality inevitably raises.

WHO AM I?

Here is a little experiment readers can try. Find several willing friends and have everyone stand in a single-file line, or in a circle, with their eyes closed and everyone close enough so that each person can reach around the one in front of them and gently rub the end of that person's nose.

Most everyone will feel as if they are touching their own nose, even though they know they're not. Their own finger movements

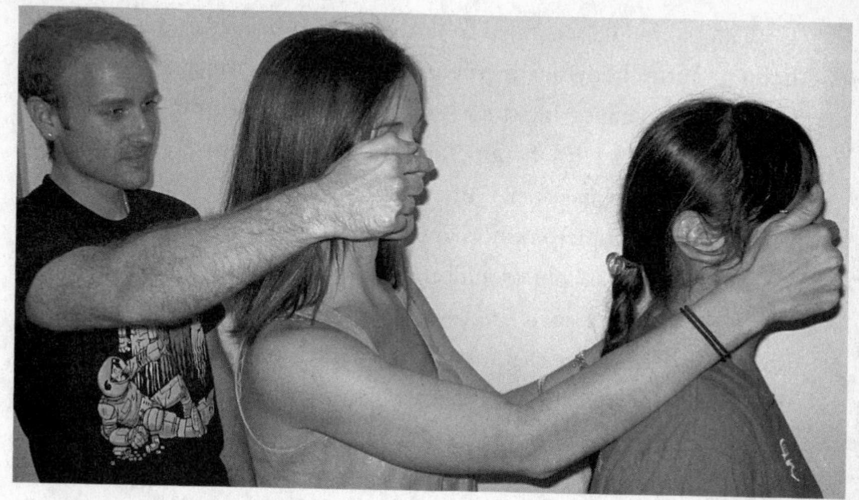

Who nose?

and the tactile sensations of rubbing the nose of the person in front of them, and the feeling of the finger of the person behind them on their own nose, overrides their knowledge that their fingers are not on their own noses. Can digital avatars be integrated into one's sense of self in an analogous manner?

Philosophers and psychologists from ancient to modern times have argued about what "self" means. When philosopher and scientist René Descartes wrote, "I think, therefore I am," he pithily described what he believed to be proof of his own existence. Moreover, he argued that the self was distinct from God, other people, and even one's own body. His argument came to be known as "Cartesian dualism." Nearly four hundred years later, António Damásio, a neuroscientist now at the University of Southern California, argued in his book *Descartes Error* that Descartes was mistaken, and that the mind is not separate from the brain and body. Today, most philosophers and scientists embrace some form of monism, arguing that mind and body are one.

Can the self be separated from the body? The idea of the human brain or entire body floating in a sea of nutrients, attached to a network of wires supporting the self somewhere else is common to both philosophy and science fiction; for example, Tufts University's Daniel Dennett's famous paper "Where Am I?" describes a brain in a vat, which controls numerous mechanical forms.

In a most basic philosophical sense, to be human is to have a range of experiences and emotions. Harvard University philosopher Robert Nozick argues that to plug into a virtual experience machine that grants unconditional pleasure would be to rob humanity of a deeper reality gained only through "actual" experiences. As we've explored, virtual life will hardly be utopian, and will bring with it the inescapable trials and tribulations inherent in human existence. But, that being said, the question remains regarding the nature of existence itself—are physical experiences somehow different on a

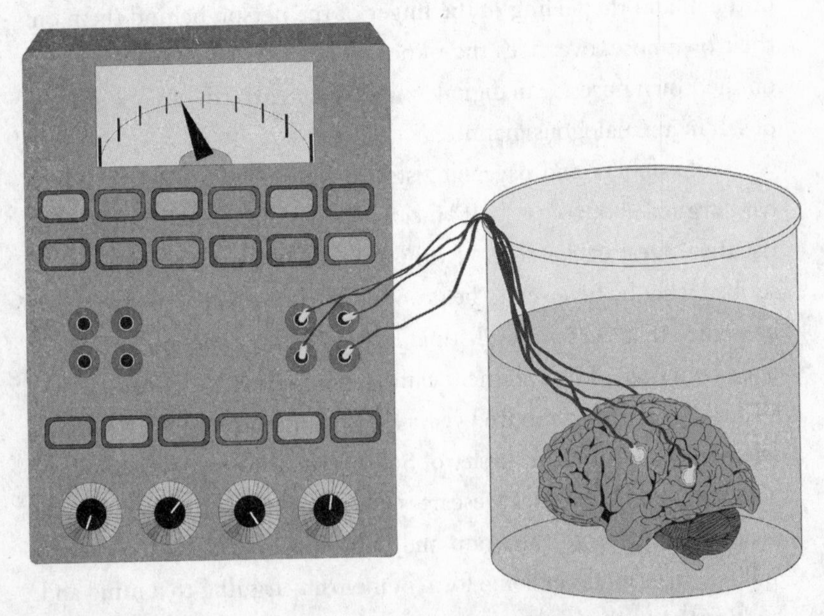

If brains could communicate directly with computers, would people need bodies?

fundamental level from virtual-reality experiences? As perceptual scientists maintain, and as we detailed at the very beginning of this book, perception is always a form of constructed information. Colors, smells, and sounds don't have a ground truth—they are experienced subjectively from one person to another, and even, from time to time, differently by the same person. So, in a sense, if all of perceived reality is virtual, then the notion that "true" experience can only be had in physical reality seems, well, unreal.

HOW MANY AM I?

Can the body be intertwined with more than one self? Shirley Ardell Mason, an only child, was born in 1923 and lived until 1998. She was an art teacher with apparent emotional problems, who ran a gallery. She reported that she would often find herself someplace where she didn't remember traveling. On occasion, she literally lost chunks of time, sometimes days. Unlike most Americans in the 1950s, Shirley entered therapy. There, her psychotherapist discovered that she had many separate identities. However, these were not the normal everyday identities associated with roles such as librarian, mother, piano player, etc. These were quite different personalities who took turns taking over her body. Shirley suffered the rare condition then labeled "multiple personality disorder." With Shirley's blessings, F. R. Schreiber wrote a book describing her life, the best-seller *Sybil.* A 1976 television miniseries of the same title, starring Sally Field, fascinated viewers.

Of course, we like to think of ourselves and others as having stable personalities. Evidence to the contrary is generally rationalized away, as in, "He's not himself today." On the other hand, people have many selves, or identities, ones appropriate to the social roles, all somewhat loosely tied together by personality and social norms appropriate to the situation. In contrast, even the roles we play are up for grabs in virtual reality. Avatars can reveal sides of people they themselves never knew existed. In *Second Life*, there are far more female avatars than there are female users. Virtual reality will raise identity switching to whole new levels. From questions of social roles, to gender, to ethnic or "tribal" affiliations—identity will become radically more plastic. Is such behavior pathological? If so, by what standards will it be judged? Scholars such as M.I.T.'s Shirley Turkle have examined this issue, and she, for one, suggests that multiple personalities in virtual reality may not only be normative but therapeutic.

Recall our discussion of the couple who met and decided to get married via a ceremony in *Second Life*. They agreed that they could enjoy cybersex extramaritally but only using different bodies, that is, avatars. The norms and ethics of behavior will continue to play out in interesting ways in the virtual age. Will traditional notions of self have any meaning? What will be the moral code? Cyberphilosophers, such as Dartmouth's James H. Moor, are examining such ethical issues, while cyber-lawyers are pondering legal issues. Rutgers law professor Greg Lastowka recently wrote in his book *Virtual Justice*, "Twenty years ago, science fiction writers imagined international communities forming within simulated online spaces . . . That future has arrived. Both law and society will need to adapt to this new environment."

ETERNAL LIFE

While medical science has increased human longevity, and stem cells may someday be regarded as the ultimate fountain of youth, the human body will remain mortal in the foreseeable future. However, as we described previously, avatars challenge fundamental notions of death.

William Sims Bainbridge wears many hats—fiction writer, avid online video-gamer, sociologist, harpsichord constructor, and, currently, program director at the National Science Foundation, where he manages computer science and engineering research. He received his Ph.D. from Harvard in the early 1970s, studying the sociology of religion. Just recently, he sent his avatar (as well as a map of his DNA) into space to reside on the international space station. Bainbridge is one of the biggest proponents of "cyber-immortality," an idea partially inspired by science fiction written in the 1950s. In his novel *The City and the Stars*, Arthur C. Clarke described how people could live forever by archiving themselves within an advanced computer.

Indeed, Bainbridge has begun to realize the vision of one of his favorite authors. He has developed numerous research projects and software applications geared toward archiving personality. One of the software packages is "The Year 2100," and can be freely downloaded. The software claims: "This is the first of a series of software modules for archiving a human personality for possible reanimation, either in a StarBase information system as a CyClone (cybernetic clone) or in a new biological body." Users who choose to archive their personality are barraged with thousands of statements relating to values, beliefs, hopes, and goals, and have to respond, rate, and elaborate upon those statements. Bainbridge is collecting a substantial database of archived digital personalities. Given what we know about people's desire to live forever, and how much more quickly digital technology is advancing compared to biological solutions to aging, virtual identity archiving will be one of the main areas of development in the coming years.

MORE HUMAN THAN HUMAN

In the American classic *Catch-22*, Joseph Heller entertains the reader with paradoxical "no-win" situations. Perhaps the most popular example involves a bombardier trying to duck combat flight duty. By regulation, the pilot cannot fly if he is "crazy." But in order to get declared crazy, he must ask for a psychological examination. By doing so, he demonstrates a rational fear of dying, hence proving he is not crazy. So, if a pilot says nothing, he flies. On the other hand, if he asks for an evaluation, he flies. Either way, he is up in the air.

In some ways, virtual reality presents a similar quandary. In this book, we illustrate two concepts at odds with each other. On the one hand, the brain treats all experiences, whether they are "real" or "virtual," the same. If it looks, sounds, feels, and smells like a person, then people treat it like another human being. On the other hand, in virtual reality, the rules of grounded reality are suspended. As Neo learns in *The Matrix*, "Do not try to bend the spoon—that's impossible. Instead, only try to realize the truth: *There is no spoon.*" In virtual reality, people have abilities that just don't exist in the physical world. Avatars can age, grow, shrink, teleport, and fly at will. They can use conversational superpowers—for example, large-scale mimicry and gaze. They can even wear other people's faces and bodies.

In sum, the brain has evolved to respond to avatars in very human ways, but avatars can be "more human than human." Is this a no-win situation? Certainly as virtual reality becomes ubiquitous, there will be winners and losers. Avatars, like any dual-use technology, will be used for good or evil. Society has always adjusted to new technology. How the world turns as avatars become more and more integrated into our daily lives will continue to fascinate us.

ACKNOWLEDGMENTS

ACADEMICS ARE QUITE CONCERNED ABOUT WHO GETS THE FIRST author credit on a publication. Each of us had good reasons to claim first authorship. Instead of debating, we did what typical scientists do in this situation: we each tried to pick the most winners one week in the middle of the 2009 NFL season. It came down to a single game. On a call later deemed a mistake by game officials, Jim won the honor.

This book is the product of a collaboration begun at the University of California, Santa Barbara, in 1998, after Jack Loomis and Jim Blascovich wrote a successful grant proposal to the National Science Foundation promoting the use of immersive virtual-reality technology to study behavior. Among other resources funded by the grant, it allowed Jim to hire a postdoctoral fellow to focus on virtual social interactions. In one of the best decisions of his career, Jim chose Jeremy Bailenson. The succeeding twelve years were special. Together with Jack Loomis and Andy Beall, we took on "virtual reality," an area that in the late nineties seemed more akin to science fiction than to science. Today virtual reality has moved from laboratories to living rooms.

The pronoun *we* is used to enhance readability throughout the book—even if a study was conducted by only one of us and even if sometimes inaccurate (for instance, Jeremy doesn't have two daughters and a son and Jim doesn't receive funding from Japanese car companies).

Three remarkable visionaries set the stage for the philosophy, ideas, and research explored in this volume:

- Jack Loomis is not only a brilliant perceptual scientist but is also a gifted intellectual, heroic aircraft pilot, thoughtful and warm colleague, and a paragon of scientific values and standards. Jack's revolutionary ideas about virtual reality were planted in the mid-eighties; the seeds that he sowed blossomed into his own formidable research programs as well as into many of the ideas, theories, and technologies described in this book. Jack invited Jim into his lab for a "pit demo." Soon after, they co-founded the Research Center for Virtual Environments and Behavior (RECVEB) at UCSB.
- Andy Beall is a modern Renaissance man: a perceptual scientist, an artist, and a technological genius. When Andy needed an affordable head-mounted display for his dissertation in the early nineties, he built one by mounting two screens cannibalized from Sony Watchman portable TVs inside a scuba mask, creating a "poor man's HMD." Little of our work would have been possible without Andy's creativity. Many of the experiments described in this book were conceived while he and Jeremy sat on surf boards at UCSB's Campus Point waiting for waves to arrive. Andy now runs perhaps the world's largest (and in our opinion, the best) comprehensive virtual reality technology company, WorldViz LLC.
- Jaron Lanier changed the course of science in the digital age. Most noted for having built some of the first virtual-

reality technology and for coining the term *virtual reality*, Jaron has been a consistent source of unique, inspirational ideas. Almost every time we thought we discovered a new take on virtual reality, it turned out that Jaron had already talked about it. His maxim that virtual reality should be a forum in which the self can undergo transformation underlies much of our work.

Scores of graduate students, postdoctoral research fellows, and undergraduate researchers have been instrumental in this work. We apologize in advance for any unintentional omissions. At UCSB, graduate students Kim Swinth, Alex Dimov, Crystal Hoyt, Susan Persky, Kim Mangel, Heidi Kane, Christine Ma, Ricardo Fonseca, and Mario Weick worked long, hard, and creatively on our various studies. Most have earned their degrees and several now have their own virtual-reality labs. Postdoctoral fellows Rosanna Guadagno, Cade McCall, Peter Kashoobeh, and Deb Bunyan followed Jeremy. All enjoy productive research careers at various institutions. Special thanks are due to former undergraduates Chris Rex, Lauren Aguilar, Tessa West, Eyal Ahroni, and Ariana Young, all well on their way in scientific careers. Our gratitude also goes to Ariana Young (again), Sara Driskell, and Natalya Lande, who served as RECVEB lab managers.

Nick Yee and Jeremy arrived at Stanford simultaneously. What followed was the most productive phase in Jeremy's career. Nick is a rare breed of scientist who can program a virtual-reality simulation from scratch and then write a paper that reads as if it were penned by a professional writer. In Jeremy's four years of doctoral training, Jeremy and Nick produced over twenty-five academic publications. Nick is currently a research scientist at the Palo Alto Research Center. Other graduate students who were a boon to the research at

Stanford are Jesse Fox, Grace Ahn, Kathryn Segovia, Maria Jabon, Hal Ersner-Hershfield, Leo Yeykelis, and Joris Janssen.

Many researchers have had notable influence on our work, either via their independent work or via direct collaboration. Thanks, in random order, to Cliff Nass, Kip Williams, Byron Reeves, Nancy Collins, Jonathan Gratch, Gary Bente, Nicole Kramer, Nikol Rummel, Rich Mayer, Ipke Wachsmuth, Fred Turner, Alex "Sandy" Pentland, Brenda Wiederhold, Ruzena Bajcsy, Phil Shaver, Howard Rheingold, Roy Pea, Dan Schwartz, Omri Gillath, Fred Brooks, Jeff Hancock, Kristine Nowak, Frank Biocca, Mel Slater, Ralph Schroeder, Mark Wiederhold, Beth Noveck, Shyam Sundar, Wijnand Ijsselsteijn, Skip Rizzo, Matthew Lombard, Elly Konijn, Sri Kalyanaraman, Shanto Iyengar, Joe Walther, Judee Burgoon, Justine Cassell, Ted Castranova, Beth Noveck, Matthew Turk, Chris Dede, Hunter Gehlbach, Scott Brave, Katrin Allmendinger, Christian Unkelbach, Greg Welch, Amy Baylor, Kwan Min Lee, and Dmitri Williams.

We thank the National Science Foundation for funding the lion's share of this work. In addition, we thank the Army Research Lab, the Mind Science Foundation, the Sage Center for the Study of the Mind, the UCSB Office of Research, the UCSB Psychology Department, the Stanford Department of Communication, the Stanford Vice Provost for Undergraduate Education, Stanford's Media-X program, the Stanford Office of Technology Licensing, the Stanford Center on Advancing Decision Making in Aging, and the National Institutes of Health.

Book writing is a complex and taxing endeavor, and we had a lot of assistance. First and foremost, our agent from William Morris Endeavor, Eric Lupfer, provided great ideas and much guidance. Our editor at William Morrow, Peter Hubbard, has been fantastic, and the book is much better for his vision and attention to detail. Thanks also to our lawyer, Lisa Digernes. A small army of undergraduate students and employees at Stanford put in much of their time on this project. Deonne Castaneda found relevant news stories. Our illustrations reflect the efforts and talents of Cody Karutz, Michelle Del Rosario, Felix Chang, Julio Mojica, Crystal Nwaneri, and Amanda Schwab. Thanks to Kathryn Segovia, Hal Ersner-Hershfeld, Tara Jones, and Jeremy Bleich for fact-checking and contributing comments on readability. Finally, thanks to Lisa Suruki, Barbara Kataoka, and Susie Ementon for managing the administrative side of the lab.

On a personal note, Jeremy thanks his mother and father, who have supported him unconditionally for the past thirty-eight years and taught him everything he knows about being a professional. It's truly a luxury to grow up without a care in the world. In addition, Myrna, Emily, Corey, Adam, Mr. and Mrs. Z, and Jim have been as wonderful a family as he could hope to have. This book would likely not be the same without Jeremy's wife, Janine, who has been instrumental in every phase of this work by reading, editing, and coming up with big ideas. In his academic life there are some special people who stand out: Woo-kyoung Ahn for inspiring him to become an academic, Doug Medin and David Uttal for their support and advising in graduate school, and of course Jim Blascovich, who went out on a limb and gave him a job when others were not so inclined. He single-handedly taught him social psychology and has been a true inspiration, epitomizing all that is wonderful about academia.

Jim thanks his gifted spouse and treasured academic colleague, Brenda Major, who has been a loving, supportive partner for more than three decades and his collaborator in raising three wonderful children, Beth, Meridith, and Greg, now adults, who are everything for which a parent hopes and more. On a professional note, many of the ideas in this book stem from seemingly unrelated ones originally seeded in his mind by his academic mentors including Patrick

Laughlin, Gerald P. Ginsburg, and Edward S. Katkin. Finally, there are no superlatives that do justice to the nature of his collaboration over the years with Jeremy Bailenson, who was not just the best post-doctoral fellow he could have imagined, but also a colleague and scientist with unparalleled drive, energy, and ideas.

NOTES

INTRODUCTION

PAGE

1 Byron Reeves and Clifford Nass report dozens of experiments that make a case that the mind has not yet evolved enough to **differentiate between real and virtual** experiences in: Reeves, B. & Nass, C. *The Media Equation: How People Treat Computers, Televisions and New Media like Real People and Places* (Cambridge, England: Cambridge University Press, 1996).

1 Several **brain imaging studies** have compared activation patterns of people interacting with real versus virtual stimuli. One study found that virtual interaction did not engage the same systems as real: Perani, D., Fazio, F., Borghese, N. A., Tettamanti, M., Ferrari, S., Decety, J. & Gilardi, M. C., "Different Brain Correlates for Watching Real and Virtual Hand Actions," *NeuroImage, 14* (2001), 749–758. However, that study used graphics that were very low in realism. A more recent study, using better graphics, noted that the brain pattern found when interacting with avatars was similar to patterns found during face-to-face interaction: Chen, D., Shohamy, V. Ross, Reeves, B., and Wagner, A., "The Impact of Social Belief on the Neurophysiology of Learning and Memory," paper presented at the Society for Neuroscience Conference, Washington, DC, 2008.

1 During an interview we conducted with Jaron Lanier (January 14, 2008), he described how he was the first to use **the term *virtual reality***, though other people had used the word *virtual* in other word pairings before.

2 In January 2010, **the Kaiser Family Foundation** released a report: Rideout, V. J., Foehr, U. G. & Roberts, D. F., *Generation M^2: Media in the Lives of 8–18 Year-olds.* It revealed that on a typical day children spend about eight hours using media. These numbers were substantially higher than they were in a similar survey taken in 2005, and are conservative, as they do not include time spent using the media for school work. The report is available at http://www.kff.org/entmedia/upload/8010.pdf (accessed September 25, 2010).

2 Data on **kids' use of video games**, print media, and movies is in the table on p. 2 of the Kaiser Family Foundation study mentioned above.

2 **Online game statistics** are from: Yee, N., "The Demographics, Motivations and Derived Experiences of Users of Massively-multiuser Online Graphical Environments," *PRESENCE: Teleoperators and Virtual Environments, 15* (2006), 309–329.

2 The first page of a leading virtual-reality textbook indicates that the **rate of virtual technology** is growing quickly: Burdea, G. & Coiffet, P. *Virtual Reality Technology*, 2nd ed. (Hoboken, New Jersey: John Wiley & Sons, 2003).

2 *Immersion* is a loaded word within the virtual-reality community. Some say that when one is immersed, she **feels as if she is inside** the medium. There are literally dozens of definitions of immersion; we have chosen one that is descriptive though not necessarily definitive.

2 ***3-D*** literally means there is a third dimension. Most computer images are 2-D, and when information is portrayed on the Z-axis, one sees depth. A desktop virtual world such as *Second Life* actually presents information on the Z-axis, but because a typical computer display is not designed to project different images to each eye corresponding to interocular distance, the effect is not one that is "stereoscopic"—for example, what happens when one wears polarized glasses at a 3-D movie. In this sense, content can be 3-D, but one can see it in a "2-D fashion," which is what occurs when one watches a 3-D movie without the glasses.

3 Philip Swann, president and publisher of TVPredictions.com, reports how **high-definition television uniquely engages felines**: http://www.tvpredictions.com/catwatcheshdtv011606.htm (accessed September 25, 2010).

3 The *New York Times* reported on January 26, 2010, that ***Avatar* had made more money** than any movie to that date in history: http://www.nytimes.com/2010/01/27/movies/awardsseason/27record.html (accessed September 25, 2010).

3 The **proportion of physically handicapped players** of avatar-based role-playing games was just over 2 percent higher than the general population, according to this study: Williams, D., Yee, N. & Caplan, S., "Who Plays, How Much, and Why? A Behavioral Player Census of Virtual World," *Journal of Computer Mediated Communication, 13* (2008), 993–1018.

4 Effects of **height and attractiveness of avatars** on the physical user are described here: Yee, N. & Bailenson, J. N., "The Proteus Effect: Self-transformations in Virtual Reality," *Human Communication Research, 33* (2007), 271–290.

4 A description of the **virtual Orville Redenbacher** advertisement appears in *USA Today:* http://www.usatoday.com/money/advertising/2007-01-11-orville-usat_x.htm (accessed September 25, 2010).

4 For example, a company named Virtual Eternity (http://www.virtualeternity.com/) provides services to **create an avatar** with clients' likenesses.

5 **Online bullying** caused a girl to commit suicide, as reported in the *Daily Telegraph* on November 20, 2007: http://www.telegraph.co.uk/news/world

news/1569949/Girl-13-commits-suicide-after-online-bullying.html (accessed September 25, 2010).

5 Julian Dibel first reported **"The Rape in Cyberspace,"** *Village Voice*, December 23, 1993.

5 A description of a U.S. government program on using **tracking data to identify personal characteristics**, "Reynard," can be found here: http://www.iarpa.gov/Reynard_BAA_Amend1.pdf (accessed September 25, 2010).

6 ***War of the Worlds,*** an adaption of H. G. Wells's novel by the same name, was broadcast on the Columbia Broadcasting System on October 30, 1938.

6 ***FarmVille*** is a game produced by Zynga. The *Independent* describes the popularity of the game on February 22, 2010: http://www.independent.co.uk/life-style/gadgets-and-tech/features/welcome-to-farmville-population-80-million-1906260.html (accessed September 25, 2010).

7 **Internet usage** statistics, including ones by country, can be found at: http://www.internetworldstats.com/stats.htm3 (accessed September 25, 2010).

7 **Corporate avatar use** is described in Reeves, B. *Total Engagement* (Cambridge, Massachusetts: Harvard Business School Press, 2009).

7 **Average growth** rate of Internet users over the last ten years has been 45 percent (http://www.internetworldstats.com/). At this rate, growth would double every two years.

CHAPTER 1. DREAM MACHINES

PAGE

9 One of the most prolific writers across genres was **Aldous Huxley**. His **perception claim** appears in Huxley, A. *The Doors of Perception* (New York: Harper and Row, 1954).

10 We are **Brussels sprouts** lovers, but our partners hate them. It turns out that taste appreciation for the vegetables is genetic: Kim, U. K. & Drayna, D., **"Genetics of Individual Differences in Bitter Taste** Perception: Lessons from the PTC Gene," *Clinical Genetics*, *67* (2004), 275.

10 We learned about **color constancy** by worrying that our local paint store had inconsistent paint. It turned out the problem wasn't the paint store. See: Maloney, L. T. & Wandell, B. D., "Color constancy: A Method for Recovering Surface Spectral Reflectance," *Journal of the Optical Society of America, 3* (1986), 29.

12 It took one of our mother-in-laws four or five viewings to actually **see the gorilla** in one of Simons's films. Chabris, C. & Simons, D. *The Invisible Gorilla and Other Ways Our Intuitions Deceive Us* (New York: Random House, 2010).

13 See: **Eberhart, J. L.**, Goff, P. A., Purdie, V. J. & Davies, P. G., "Seeing Black: Race, Crime, and Visual Processing," *Journal of Personality and Social Psychology*, *87* (2004), 876. This research suggests that getting pulled over for **driving while black** may be reinforced by the enhanced ability of police to spot weapons when primed by an African American face.

16 The "**Müller-Lyer Illusion**" was developed by F. C. Müller-Lyer in the late 1800s.

16 This ingenious **prism-glasses** self-study can be found at: Stratton, G. G., "Vision Without Inversion of the Retinal Image," *Psychological Review*, 4 (1897), 341.

16 Anatomically, all **visual sensations** are normally inverted on the human retina to begin with, so the prism glasses actually focused them right-side-up on the retina, righting the usually upside-down perception. Got that? So the situation is even more complicated—and relative—than it might first appear.

19 Do we really let our **minds wander** two thousand times a day on average? To find out, read Schooler, J. & Smallwood, J. W., "The Restless Mind," *Psychological Bulletin, 132* (2006), 946; or Klinger, E., "Daydreaming and Fantasizing: Thought Flow and Motivation," in Markman, K. D., Klein, W. M. P. & Sahr, J. A. (eds.), *Handbook of Imagination and Mental Stimulation* (New York: Psychology Press, 2009), 225–240.

19 The exhaustion and mental confusion that nag people suffering from sleep apnea are often attributed to lack of sleep itself rather than a lack of dreaming, but the latter contributes. The National Institute of Health provides many types of **data on sleep apnea** here: http://www.ninds.nih.gov/disorders/brain_basics/understanding_sleep.htm#dreaming9 (accessed September 25, 2010).

21 Of course, many types of **drugs can be addictive**, see: Haertzen, C. A., Kocher, T. R., and Miyasato, K., "Reinforcements from the First Drug Experience Can Predict Later Drug Habits and/or Addiction: Results with Coffee, Cigarettes, Alcohol, Barbiturates, Minor and Major Tranquilizers, Stimulants, Marijuana, Hallucinogens, Heroin, Opiates and Cocaine," *Drug and Alcohol Dependence, 11* (1983), 147–165.

22 Generic names for **Stelazine** and **Thorazine** are trifluoperazine and chlorpromazine, respectively.

CHAPTER 2. A MUSEUM OF VIRTUAL MEDIA

PAGE

25 These interpretations are based on perusals of Web sites devoted to the **Lascaux paintings**, the most authoritative of which is probably http://www.lascaux.culture.fr/#/fr/00.xml (accessed September 25, 2010), available in several languages, including English. For a discussion, see: Curtis, G. *The Cave Painters: Probing the Mysteries of the World's First Artists* (New York: Anchor Books, 2006), and Knecht, H., Pike-Tay, A. & White, R. (eds.), *Before Lascaux: The Complex Record of the Early Upper Paleolithic* (Boca Raton, Florida: CRC Press, 1993).

27 A debate about the **bleeding religious icons** is here: http://www.religioustolerance.org/chr_stat.htm (accessed September 25, 2010).

27 Never thought **ink** had a long and complicated history? See: http://inventors.about.com/library/weekly/aa100197.htm (accessed September 25, 2010).

28 The **Bible** may not have been the first-ever printed text, but it is the best-

known: Kapr, A. (Martin, Douglas, transl.). *Johann Gutenberg: The Man and His Invention* (Aldershot, England: Scolar Press, 1996), 3rd ed., revised by the author for the English translation from the German.

28 **Harry Potter** sales estimates are based on interpolation from the story reported by http://www.theBookseller.com in June 2008 (accessed September 25, 2010).

29 The "**camera obscura**" uses "pin-hole" projection onto a surface that could then be traced in ink or other medium to save photo-realistic images. See a description here: Crombie, Alistair Cameron. *Science, Optics, and Music in Medieval and Early Modern Thought* (London, England: Continuum International Publishing Group, 1990).

For a clearly written, authoritative history of photography, examine Marien, M. W. *Photography: A Cultural History*, 2nd ed. (London: Laurence King Publishing, 2006).

30 **Color photography** was not Maxwell's only contribution. See: Mahon, B. *The Man Who Changed Everything—the Life of James Clerk Maxwell* (Hoboken, New Jersey: John Wiley & Sons, 2003).

31 The "Dark Lady of American Letters" reviews the history of the ***Ciotat* movie**. See: Sontag, Susan, "The Decay of Cinema," *New York Times Magazine*, February 26, 1996.

32 The **Pony Express** disappeared after about eighteen months. Instead of a week, message transmission became a matter of seconds, quite an increase in what we today would call "processing speed." See: Visscher, V. L. *A Thrilling and Truthful History of the Pony Express* (Chicago: Rand McNally & Co., 1908).

33 Who cares about the **vacuum tube**? We should all appreciate it. See: Harr, C., "Ambrose J. Fleming Biography. Pioneers of Computing. The History of Computing Project," http://www.thocp.net/biographies/fleming_ambrose.htm (accessed September 25, 2010).

33 **ENIAC** was transferred to the U.S. military's Aberdeen proving ground in Maryland in 1947, where it operated continuously until 1955. See: Norman, Jeremy M., ed., *From Gutenberg to the Internet: A Sourcebook on the History of Information Technology* (Novato, California: Historyofscience.com, 2005).

33 The forerunner of such higher-level networking was the **ARPANET**, an "internet" created by the U.S. military. In 1969, it was permanently established with four nodes—UCLA, UCSB, Stanford Research Institute, and University of Utah. See: Rheingold, H. *The Virtual Community: Homesteading on the Electronic Frontier* (New York: Harper Perennial, 1993). Go Cardinal and Gauchos!

34 Although **Vice President Gore** never actually said that he "invented" the Internet, political opponents put the words in his mouth during the 2000 U.S. presidential campaign. For a description of the controversy, see: "Did Gore Invent the Internet?," October 5, 2000, http://www.salon.com/technology/col/rose/2000/10/05/goreinternet (accessed September 25, 2010).

34 If it is difficult to believe the **number of Web sites** in 1992 was so low, then consider the very low number of televisions in American homes in

1948. See: http://royal.pingdom.com/2008/04/04/how-we-got-from-1-to-162-million-websites-on-the-internet/ (accessed September 25, 2010).

CHAPTER 3. MIRROR, MIRROR ON THE WALL

PAGE

39 **Jack Loomis** has been involved either directly or indirectly in a majority of the virtual-reality experiments we have run. We consider him to be one of the major true pioneers studying virtual reality, especially in terms of how the human perceptual system interacts with "virtual" and "real" objects. Jack's 1992 paper on how and why people experience presence in virtual reality is a masterpiece: Loomis, J. M., "Distal Attribution and Presence," *Presence: Teleoperators and Virtual Environments*, *1* (1992), 113–119. During an interview with us, he described an early meeting of the soon-to-become players in the field of virtual reality and "presence" research, including himself, Jaron Lanier, and Nat Durlach, among others, at the Santa Barbara Sheraton, sometime around 1990.

41 **Fred Brooks** has forgotten more about virtual-reality research than we will ever know. We had the fantastic experience of interviewing him over dinner while he described his research history in the field. One paper of his that is particularly influential to many in the field is: Brooks, F. P. Jr., "What's Real About Virtual Reality?," *IEEE Computer Graphics and Applications*, *19* (1999), 6: 16–27.

43 The tale of the ***Pong* arcade game** that broke down from overuse is described by Scott Cohen in his book *ZAP! The Rise and Fall of Atari* (New York: McGraw-Hill, 1984).

48 An article in *Time* magazine on April 5, 1982, describes the cultural phenomenon known as **"*Pac-Man* fever"**: http://www.time.com/time/magazine/article/0,9171,921174,00.html (accessed September 25, 2010).

48 **Jaron Lanier** is perhaps the name most associated with "virtual reality." In addition to coining the term, he built the first color head-mounted display and probably designed the first systems in which avatars were tracked and rendered. At the age of twenty, he formed VPL, a company specializing in virtual-reality technology. Today he remains very active as a thought leader in the area. His metaphor of **perceptual system as a spy submarine** comes from our interview with him.

48 For a description of how **eye movements facilitate seeing**, see: Findlay, J. M., and Gilchrist, I. D. *Active Vision—The Psychology of Looking and Seeing* (Oxford, England: Oxford University Press, 2003).

49 An article by Tracy Clark-Flory titled "Grand Theft Misogyny" (May 3, 2008) describes the graphic process of paying for sex and then **killing the prostitute to get the money back** in the video game *Grand Theft Auto:* http://www.salon.com/life/broadsheet/2008/05/03/gta (accessed September 25, 2010)

51 **Ivan Sutherland**'s most referenced paper on this topic is: Sutherland, I. E., "The Ultimate Display," invited lecture, IFIP Congress 65. An abstract ap-

pears in *Information Processing 1965: Proc. IFIP Congress 65*, vol. 2, Kalenich, W. A., ed. (Washington, DC: Spartan Books, and New York: Macmillan), 506–508.

54 If you're curious, the **Computer History Museum** is just off Highway 101 and tucked between a movie theater complex and a 7-Eleven. It presents an impressive archive of pioneering gadgets.

55 Sutherland's head-mounted display was labeled the **"Sword of Damocles"** because it hung from the ceiling as the sword from the fable did. One can also make the inference that its moniker was due to how heavy and large the display was. It was dangerous, like a sword, as well.

55 Descriptions of the **Sensorama** demonstrations come from our interview with Jaron Lanier.

57 A description of the **super cockpit** can be found in Furness, T., "The Super Cockpit and Its Human Factors Challenges," *Proceedings of the Human Factors Society, 30* (1986), 48–52. Quote from Tom Furness comes from Cipalla, Rita, "The Brave New World of Virtual Reality," *Smithsonian News Service*, December/January 2000. In addition to Furness, NASA scientist Steve Ellis has been working in this area for more than twenty years and his contributions are noteworthy.

57 For a thorough description of early **CAVE** technology, see: Cruz-Neira, Carolina, Sandin, Daniel J., DeFanti, Thomas A., Kenyon, Robert V., and Hart, John C., "The CAVE: Audio Visual Experience Automatic Virtual Environment," *Communications of the ACM, 35*, 6 (1992), 64–72

60 For a thorough discussion of the **difference between agents and avatars**, see: Bailenson, J. N. & Blascovich, J., "Avatars," *Encyclopedia of Human-Computer Interaction* (Great Barrington, Massachusetts: Berkshire Publishing Group, 2004), 64–68.

60 In this paper, published by Chip Morningstar and Randy Farmer two years before *Snow Crash* was published, **the term *avatar*** is used to describe virtual representations: Morningstar, C., and Farmer, F. R., "The Lessons of Lucasfilm's Habitat," First International Conference on Cyberspace, Austin, Texas, 1990.

61 Griffin, J. H., ***Black Like Me*** (Boston: Houghton Mifflin, 1961).

CHAPTER 4. WINNING VIRTUAL FRIENDS AND INFLUENCING VIRTUAL PEOPLE

PAGE

67 Kip Williams is actually quite a sociable fellow. To find **Williams's personal ostracism anecdote** in its original context, see Williams, K. D., "Ostracism: The Kiss of Social Death," *Social and Personality Compass, 1* (2007), 236–247. Also see: Williams, K. D. & Jarvis, B., "Cyberball: A Program for Use in Research on Interpersonal Ostracism and Acceptance," *Behavior Research Methods, 38* (2006), 174–180, and Eisenberger, N. I., Lieberman, M. D. & Williams, K. D., "Does Rejection Hurt? An fMRI Study of Social Exclusion," *Science, 302* (2003), 290–292.

67 Williams and his colleagues have run many different experiments using ***Cy-***

berball. Indeed, more than five thousand participants have played online. This online ostracism with cartoonish avatars produced profound negative reactions. These effects are so robust that when the researchers go out of their way to reduce them, the responses persist. For example, even if participants lose money every time the digital ball is tossed to them, they still feel ostracized when they are left out of the game. Furthermore, it doesn't matter if participants believe that the other two players are friends or strangers, if they belong to the participants' team or not, if they are similar in socioeconomic status to the participants or not, or even if they are perceived as despicable people or not.

Williams and his colleagues conducted a *Cyberball* experiment where they manipulated the supposed groups to which the other two players belonged. In one condition, the other two players were portrayed as "in-group" members—specifically, members of the same Australian political party as the participant. In another condition, the other players were portrayed as rival "out-group" members—specifically, members of a different Australian political party than the participant. In the third condition, the other players were portrayed as "despised out-group" members—specifically, members of the Australian wing of the Ku Klux Klan (KKK). Despite the fact that the participants in the latter condition disagreed with the beliefs of the KKK, didn't respect its members, were disgusted by the group, and believed the world would be a better place without them, they were still as negatively affected by their ostracism experience as participants both in the in-group and out-group conditions.

68 Most medical **MRI scans** are performed to search for anatomical abnormalities such as tumors, blood clots, or aneurysms. Although medical personnel, such as radiologists, use the same type of MRI machine, social cognitive neuroscientists conduct brain scans to locate brain activity rather than abnormalities. More specifically, they look for the locations of increases or decreases in blood flow in the brain that co-occur during the thoughts, feelings, and/or actions relevant to the social psychological processes in which they are interested. The latter technique is called functional magnetic resonance imaging, or "fMRI." Nearly all MRI scans, whether structural or functional, are noninvasive. The person being scanned simply lies still in a tube within a rather large machine while the machine does its thing (though it is a claustrophobic, noisy experience).

69 To learn more about **social psychology,** see: Fiske, S. T., Gilbert, D. T., and Lindzey, G. (eds.), *Handbook of Social Psychology,* 5th ed. (Hoboken, New Jersey: John Wiley & Sons, 2010).

70 See Epley, N., Waytz, A., Akalis, S. & Cacioppo, J. T., "When We Need a Human: Motivational Determinants of **Anthropomorphism**," *Social Cognition, 26* (2008), 143–145.

70 See the classic book written by our good friends and colleagues Reeves, B. & Nass, C. ***The Media Equation**: How People Treat Television, New Media and Computers* (Wayne, Pennsylvania: CLSI Publications, 1996).

71 Seen **"Clippy"** lately? See: Swartz, L. *Why People Hate the Paperclip: Appear-*

ance, Behavior, and Social Responses to User Interface Agents (Honors thesis, Symbolic Systems, Stanford University, 2003).

73 Scientists like to explain *why* something happens. Why do people get so immersed in daydreams, books, movies, and online games full of fictional characters? How is it that humans interact with digital "beings" as if such computer creations are actually people? To answer questions like these, scientists develop theories to discover, explain, and even control the links between causes and effects. In its simplest form, **a scientific theory** can be represented by the formula "A causes B." For example, gravity causes the moon to stay in its orbit around the Earth, cigarette-smoking causes cancer, or frustration causes aggression. Additionally, scientists want to be as precise and detailed as possible in explaining the linkages between causes and effects. This precision is achieved by testing and verification. Theoretical predictions—hypotheses—are tested on the basis of observable data, resulting in acceptance or rejection of them. Behavioral scientists have developed a variety of specific theories that explain the way humans relate to one another in social situations—ones such as conformity, persuasion, prejudice, and racism, to name a few. Our theory of social influence was initially described in: Blascovich, J., Loomis, J., Beall, A. C., Swinth, K., Hoyt, C., and Bailenson, J. N. "Immersive Virtual Environment Technology as a Methodological Tool for Social Psychology," *Psychological Inquiry, 13* (2002), 103–124.

74 **Terri Schiavo** was the subject of a conflict between her parents and her ex-husband regarding whether she would ever recover any cognitive function from a prior brain injury that was sustained while she and the latter were married. Her facial movements were taken as evidence of cognitive function by her parents, but not by medical personnel and her ex. The story became a moral dilemma that fed political debate in 2007. See: Goodnough, Abby, "Husband Takes Schiavo Fight Back to Politicians," http://www.nytimes.com/2006/08/16/washington/16schiavo.html?_r=1&ref=terri_schiavo (accessed September 25, 2010).

75 **Social stigma** was first studied in depth by sociologist Erving Goffman. See: Goffman, E. *Asylums* (New York: Penguin Books, 1968).

75 Data on the **relative proportions of verbal and nonverbal communication** can be found here: Burgoon, J. K., Buller, D. B. & Woodall, W. G. *Nonverbal Communication: The Unspoken Dialogue* (New York: HarperCollins/Greyden Press, 1989).

76 **High definition** probably does contribute to a more immersive visual experience for viewers, even without photorealism, e.g., Buzz Lightyear.

76 The design of **nonverbal behavior animation** is a large research area. Three notable works are: Badler, Norman, Phillips, Cary B., and Webber, Bonnie Lynn. *Simulating Humans: Computer Graphics, Animation and Control* (Oxford, England: Oxford University Press, 1993); Cassell, Justine. *Embodied Conversational Agents* (Cambridge, Massachusetts: M.I.T. Press, 2000); and Gratch, J., Rickel, J., Andrea, E., Cassell, J., Petajan, E., and Badler, N., "Creating Interactive Virtual Humans: Some Assembly Required," *IEEE Intelligent Systems* (July–August 2002), 54–63.

77 The **realism meta-analytic data** are here: Yee, N., Bailenson, J. N., Rickertsen, K., "A Meta-analysis of the Impact of the Inclusion and Realism of Human-like Faces on User Experiences," *Proceedings of CHI*, 2007 (New York: ACM Press, 2007), 1–10. For more on realism, see: Nowak, K. & Biocco, F., "The Effect of Agency and Anthropomorphism on Users' Sense of Telepresence, Copresence, and Social Presence in Virtual Environments," *Presence: Teleoperators and Virtual Environments, 12* (2003), 481–494.

78 **What's uncanny about the valley?** See: Mori, M., "*Bukimi no tani* [The Uncanny Valley]," *Energy*, 7, 4 (1970), 33–35.

79 It is worth examining how **theory of mind and communicative realism** jointly influence social behaviors in virtual reality. Imagine a character in virtual reality that behaved only slightly like a human. In other words, it has some gestures; for example, head movements, but those gestures are very limited. We can measure how realistically people behave toward this virtual human—for example, how much they look at it, how much they respect its personal space, and how polite they are to it. It turns out that if one thinks that this virtual human is an agent controlled by a computer (perhaps because someone informed them in advance), she would not treat it like a person. On the other hand, if one believes it is an avatar, controlled by another human being in real time, then despite the low realism, she would interact with it more realistically. In other words, avatars do not require high levels of realism to be effective in an interaction.

Now imagine the same scenario but the virtual human behaves extremely realistically, exhibiting gaze behavior, walking normally, and moving its hands. In this case, despite the fact that it is known to be controlled by a computer, even an agent can elicit realistic social and emotional responses. The bottom line is that avatars can influence people regardless of how realistically they behave, but agents influence people only to the extent that they exhibit high communicative realism.

79 Gladwell popularized **unconscious** decisions in his best-seller: Gladwell, M. *Blink: The Power of Thinking Without Thinking* (New York: Little, Brown and Company, 2005).

CHAPTER 5. THE VIRTUAL LABORATORY

PAGE

84 Read: Crocker, J., Major, B., and Steele, C., "**Social Stigma,**" in Gilbert, D. T., Fiske, S. & Lindzey, G. (eds.), *Handbook of Social Psychology*, 4th ed. (Boston, Massachusetts: McGraw-Hill, 1998), vol. 2, 504–553.

85 We reported the **cardiovascular effects** of interacting with socially stigmatized individuals in: Blascovich, J., Mendes, W. B., Hunter, S. B., Lickel, B. & Kowai-Bell, N., "Perceived Threat in Social Interactions with Stigmatized Others," *Journal of Personality and Social Psychology, 80* (2001), 753–767.

85 The results of our **"Sally"** experiment are here: Blascovich, J., "(Virtual) Reality, Consciousness, and Free Will," in Baumeister, R. & Vohs, K. (eds.),

Free Will and Consciousness: How Might They Work? (New York: Oxford University Press, 2010), 172–190.

86 As virtual-reality researchers, we owe a great debt to **E. T. Hall.** Proxemic measures are quite informative, sensitive, and easier to measure in virtual reality—via the tracking that occurs anyway—than in physical reality. See: Hall, E. T., *The Hidden Dimension* (Garden City, New York: Doubleday, 1966).

87 For discussion and evidence of **deviants' proxemic behaviors,** see: Felipe, N. J. & Sommer, R., "Invasions of Personal Space," *Social Problems, 14* (1966), 206–214.

88 Those interested in the details of our **simple personal-space study** are pointed to: Bailenson, J., Blascovich, J., Beall, A. C. & Loomis, J. M., "Interpersonal Distance in Immersive Virtual Environments," *Personality and Social Psychology Bulletin, 29* (2003), 819.

88 Readers can learn more about **gender differences** in proxemic behaviors here: Fisher, J. D. & Byrne, D., "Too Close for Comfort: Sex Differences in Response to Invasions of Personal Space," *Journal of Personality and Social Psychology, 32* (1975), 15–21.

89 This study required spending much time in **casinos,** but someone had to do it. Blascovich, J., Ginsburg, G. P. & Howe, R. C., "Blackjack, Choice Shifts in the Field," *Sociometry, 39* (1976), 274–276. Our first study in the **virtual casino** was reported by: Swinth, K. & Blascovich, J., "Conformity to Group Norms in an Immersive Virtual Environment," paper presented at the Annual Meeting of the American Psychological Society, Toronto, Ontario, Canada, 2001. The **real money** study is reported in: Blascovich, J., Ginsburg, G. P. & Howe, R. C., "Blackjack and the Risky Shift II: Monetary Stakes," *Journal of Experimental Social Psychology, 11* (1975), 224–232.

91 The **earliest experiment** of this kind in history is reported in: Triplett, N., "The Dynamogenic Factors in Pacemaking and Competition," *American Journal of Psychology, 9* (1898), 507–533. We reported the earliest virtual **social facilitation/inhibition** study 105 years later in: Hoyt, C., Blascovich, J. & Swinth, K., "Social Inhibition in Immersive Virtual Environments," *Presence, 12* (2003), 183.

93 This groundbreaking work on **mimicry** can be found at: Chartrand, T. & Bargh, J. "The Chameleon Effect: The Perception-behavior Link and Social Interaction," *Journal of Personality and Social Psychology, 76* (1999), 893. The wizards of digital virtual mimicry reported their work in: Bailenson, J. B. & Yee, N., "Digital Chameleons: Automatic Assimilation of Nonverbal Gestures in Immersive Virtual Environments," *Psychological Science, 16* (2006), 814.

CHAPTER 6. WHO AM I?

PAGE

96 For a description of the **agent Max**, see: Wachsmuth, I., "'I, Max'—Communicating with an Artificial Agent," in Wachsmuth, I. & Knoblich, G.

(eds.), *Modeling Communication with Robots and Virtual Humans* (Berlin: Springer, 2008), 279–295.

97 The debate over the relative influence of genetic and environmental components of **the self** is discussed in: Plomin, R., *Genetics and Experience: The Interplay Between Nature and Nurture* (Thousand Oaks, California: Sage, 1994).

98 **"When you're smiling"** by Larry Shay, Mark Goodwin, and Joe Goodwin.

99 The **facial feedback hypothesis** is described in: Zajonc, R. B., Murphy, S. T. & Inglehart, M., "Feeling and Facial Efference: Implications for the Vascular Theory of Emotion," *Psychological Review, 96* (3) (1989), 395–416.

99 The **smiling studies** are reported in: Strack, F., Martin, L. & Stepper, S., "Inhibiting and Facilitating Conditions of the Human Smile: A Nonobtrusive Test of the Facial Feedback Hypothesis," *Journal of Personality and Social Psychology, 54* (1988), 768–777.

99 For a description of **facial muscles and affect**, see: Cacioppo, J. T., Petty, R. P., Losch, M. E. & Kim, H. S., "Electromyographic Activity over Facial Muscle Regions Can Differentiate the Valence and Intensity of Affective Reactions," *Journal of Personality and Social Psychology, 50* (1986), 260–268.

100 **Self-perception theory** is detailed in: Bem, D. J., "Self-perception Theory," in Berkowitz, L. (ed.), *Advances in Experimental Social Psychology* (New York: Academic Press, 1972), vol. 6, 1–62.

100 The **heartbeat/centerfold** study is described in: Valins, S., "Cognitive Effects of False Heart-Rate Feedback," *Journal of Personality and Social Psychology, 4* (1966), 400–408.

101 For Oakland Raiders' fans, the Cornell studies on **uniform color and aggressiveness** are described here: Frank, M. & Gilovich, T., "The Dark Side of Self and Social Perception: Black Uniforms and Aggression in Professional Sports," *Journal of Personality and Social Psychology, 54* (1988), 74–85.

101 Jeff Hancock and his colleagues replicated Gilovich's NFL **work on black uniforms**. These researchers found that people whose avatars wore dark clothes behaved more aggressively than those whose avatars wore light clothes. They point out that avatars allow users to be completely anonymous, and when people have their true identities hidden, they tend to rely more on social categories and stereotypes, which can easily be cued by uniforms. The study is described here: Merola, N., Penas, J. & Hancock, J., "Avatar Color and Social Identity Effects: On Attitudes and Group Dynamics in Virtual Realities," paper presented at the ICA 2006, Dresden, Germany.

102 The **nurse/KKK** study is reported by: Johnson, R. & Downing, L., "Deindividuation and Valence of Cues: Effects on Prosocial and Antisocial Behavior," *Journal of Personality and Social Psychology, 37* (1979), 1532–1538.

103 The cited work by **Sherry Turkle** is described in: *Life on the Screen: Identity in the Age of the Internet* (New York: Simon & Schuster, 1995) and Turkle, S., "Who Am We?" *Wired, 4* (January 1996).

105 The **online dating and height** study is reported by: Hitsch, G. J., Hortacsu, A. & Ariely, D. (in press), "What Makes You Click?—Mate Preferences in Online Dating," *Quantitative Marketing and Economics.*

105 Yee isolated **self-perception** from the perceptions of the other actors in virtual reality. The others' avatars may have treated the taller or shorter participant differently as a function of its height. This differential treatment may, in turn, have actually altered the participant's behavior. To prevent such an artifact from occurring, the other negotiator, who was actually an experimental "confederate" and who always used exactly the same negotiating strategy, regardless of experimental condition, was always kept "blind" to the experimental condition—regardless of whether the first participant's avatar was taller or shorter, the second one always saw the avatar at the participant's actual height. In this manner, Yee was able to isolate purely the effect of self-perception.

106 Data showing that **height and attractiveness of avatars** change behavior in virtual reality and also carry over to the physical world are reported by: Yee, N., Bailenson, J. N. & Ducheneaut, N., "The Proteus Effect: Implications of Transformed Digital Self-representation on Online and Offline Behavior," *Communication Research, 36* (2) (2009), 285–312.

107 The Hancock paper on **deception in dating** profiles is: Hancock, J. T., Toma, C. & Ellison, N., "The Truth About Lying in Online Dating Profiles," *Proceedings of the CHI* (New York: ACM Press, 2007), 449–452.

108 The follow-up data on **avatar attractiveness** are here: Yee, N. & Bailenson, J. N., "The Difference Between Being and Seeing: The Relative Contribution of Self-perception and Priming to Behavioral Changes via Digital Self-representation," *Media Psychology, 12* (2) (2009), 195–209.

CHAPTER 7. RE-CREATING YOURSELF

PAGE

110 The **virtual aging** study appears in: Ersner-Hershfield, H., Bailenson, J. N. & Carstensen, L. L., "Feeling More Connected to Your Future Self: Using Immersive Virtual Reality to Increase Retirement Saving," poster presented at the Association for Psychological Science Annual Convention, Chicago, 2008.

111 Gibson, W., ***Neuromancer*** (New York: Ace Books, 1984).

111 A description of the data collection methodology in the *Second Life* **obesity study** can be found here: Harris, H., Bailenson, J. N., Nielsen A. & Yee, N., "The Evolution of Social Behavior over Time in *Second Life*," *PRESENCE: Teleoperators & Virtual Environments, 18* (6) (2009), 294–303.

113 The **avatar race study** is reported in: Groom, V., Bailenson, J. N. & Nass, C., "The Influence of Racial Embodiment on Racial Bias in Immersive Virtual Environments," *Social Influence, 4* (1) (2009), 1–18.

114 See Kalyanaraman, S. S., Penn, D. L., Ivory, J. D., Judge, A., "The **Virtual Doppelgänger**: Effects of a Virtual Reality Simulator on Perceptions of Schizophrenia," *Journal of Nervous and Mental Disease, 198* (2010), 437–443.

115 All of the studies on **doppelgängers** are described in this chapter: Bailenson, J. N. & Segovia, K. Y., "Virtual Doppelgängers: Psychological Effects of Avatars Who Ignore Their Owners," in Bainbridge, W. S. (ed.),

Online worlds: Convergence of the real and the virtual (New York: Springer, 2010), 175–186.

115 Description of the **Bobo Doll** study and **Social Learning Theory** can be read here: Bandura, A., *Social Learning Theory* (Englewood Cliffs, New Jersey: Prentice Hall, 1977).

117 For a discussion of **false memory** formation, see: Loftus, E. F. & Pickrell, J. E., "The Formation of False Memories," *Psychiatric Annals, 25* (12) (1995), 720.

120 The white paper describing the use of the virtual mirror for **soldiers in Iraq** is an unpublished manuscript that was produced for the U.S. Army. It is non-classified.

CHAPTER 8. STREET SMARTS

PAGE

122 For a discussion of how virtual reality can improve social interaction, see: Walther, J. B., "Computer-Mediated Communication: Impersonal, Interpersonal, and Hyperpersonal Interaction," *Communication Research, 23* (1996), 3–43.

125 Wondering about **emotional intelligence?** Read: Salovey, P. & Mayer, J. D., "Emotional Intelligence," *Imagination, Cognition, and Personality, 9*, (1990) 185–211.

125 Sternberg's theory about **"street smarts,"** as well as the other components of intelligence, is described in: Sternberg, R. J. *Beyond IQ: A Triarchic Theory of Intelligence* (Cambridge, England: Cambridge University Press, 1985).

127 The study on **non-zero sum gaze** is described here: Bailenson, J. N., Beall, A. C., Blascovich, J., Loomis, J. & Turk, M., "Transformed Social Interaction, Augmented Gaze, and Social Influence in Immersive Virtual Environments," *Human Communication Research, 31* (2005), 511–537.

128 The experiment on **audience conformity** can be found by accessing: McCall, C., Bailenson, J. N., Blascovich, J. & Beall, A. C., "Leveraging Collaborative Virtual Environment Technology for Inter-population Research on Persuasion in a Classroom Setting," *PRESENCE: Teleoperators & Virtual Environments, 18* (5) (2009), 361–369.

129 The study using avatar translucency to improve **teacher gaze behavior** appears in: Bailenson, J. N., Yee, N., Blascovich, J., Beall, A. C., Lundblad, N. & Jin, M., "The Use of Immersive Virtual Reality in the Learning Sciences: Digital Transformations of Teachers, Students, and Social Context," *Journal of the Learning Sciences, 17* (2008), 102–141. The follow-up study on autism is not yet published at the time of this writing, but was funded by the National Institutes of Health.

131 The negative response to **Hillary Clinton mimicking** a Southern drawl was reported by Salon.com on March 14, 2007: http://www.salon.com/news/opinion/camille_paglia/2007/03/14/coulter (accessed September 25, 2010).

132 The experiment showing that **agents who mimic** the head movements of users are more effective than other agents is described in: Bailenson, J. N.

& Yee, N. "Digital Chameleons: Automatic Assimilation of Nonverbal Gestures in Immersive Virtual environments," *Psychological Science, 16* (2005), 814–819.

133 Bailenson, J. N. & Yee, N., "Virtual Interpersonal Touch and Digital Chameleons," *Journal of Nonverbal Behavior, 31* (2007), 225–242, reports the **hand-shaking** study.

133 According to his Web site, Kevin Eikenberry "is an expert in converting organizational, team and individual potential into desired results," and the "Chief Potential Officer of The Kevin Eikenberry Group." He provides guidelines on the **best way to shake hands** in his book on leadership: Eikenberry, K. *Remarkable Leadership: Unleashing Your Leadership Potential One Skill at a Time* (San Francisco: Jossey-Bass, 2010).

134 The study demonstrating the negative **consequences of detecting mimicry** is reported in: Bailenson, J. N., Yee, N., Patel, K. & Beall, A. C., "Detecting Digital Chameleons," *Computers in Human Behavior, 24* (2007), 66–87.

136 The two studies using Ruzena Bajcsy's Berkeley lab to study tai chi learning with **avatars that share body space** with virtual teacher agents are reported in: Bailenson, J. N., Patel, K., Nielsen, A., Bajcsy, R., Jung, S. & Kurillo, G., "The Effect of Interactivity on Learning Physical Actions in Virtual Reality," *Media Psychology, 11* (2008), 354–376.

CHAPTER 9. ETERNAL LIFE

PAGE

138 For some important but depressing thoughts about **death,** read: Becker, Ernest. *The Denial of Death* (New York: Simon & Schuster, 1973).

138 For a historical account of the fable, see: Damrosch, David. *The Buried Book: The Loss and Rediscovery of the Great **Epic of Gilgamesh*** (New York: Henry Holt and Company, 2007).

139 A scientific discussion of the relationships between **telomeres and aging** is: Aubert, G., and Lansdorp, P. M., "Telomeres and Aging," *Physiological Reviews 88* (2) (April 2008), 557–579.

140 A thorough discussion of underlying **brain structure** is: Gazzaniga, M. S. *Human: The Science Behind What Makes Us Unique* (New York: HarperCollins, 2008).

140 On October 2, 2009, the *New York Daily News* describes how technicians hit **Ted Williams's frozen head** with a monkey wrench: http://www.nydailynews.com/news/national/2009/10/02/2009-10-02_book_reveals_chilling_details_of_how_cryonic_lab_thumped_remains_of_baseball_imm.html (accessed September 25, 2010).

141 **Ray Kurzweil** provides a compelling look into the future, arguing that digital immortality is possible, in Kurzweil, Raymond, *The Singularity Is Near* (New York: Viking, 2005).

142 **Terror management theory** is described in this paper: Greenberg, J., Pyszczynski, T. & Solomon, S., "The Causes and Consequences of the

Need for Self-esteem: A Terror Management Theory," in Baumeister, R. E. (ed.), *Public Self and Private Self* (New York: Springer-Verlag, 1986), 189–212.

143 The logic for and data supporting the idea that mortality salience causes support for **incumbent candidates** is here: Vail, K. E., Arndt, J., Motyl, M. & Pyszczynski, T., "Compassionate Values and Presidential Politics: Mortality Salience, Compassionate Values and Support for Barack Obama and John McCain in the 2008 Presidential Election," *Analyses of Social Issues and Public Policy, 9* (2009), 255–268.

147 **The lawsuit** against Electronic Arts is described in the *Washington Post*: http://www.washingtonpost.com/wp-dyn/content/article/2010/07/11/AR2010071103062.html (accessed September 24, 2010).

147 On December 26, 2008, the *Daily Telegraph* reported on how **John Lennon** appeared in an advertisement long after his own death: http://www.telegraph.co.uk/technology/3966114/Beatles-legend-John-Lennon-makes-charity-television-commercial-28-years-after-death.html (accessed September 25, 2010).

148 The article about **digitizing old actors** was published in the *Los Angeles Times* on August 9, 1999: http://articles.latimes.com/1999/aug/09/business/fi-64043 (accessed September 24, 2010).

149 The National Science Foundation–sponsored project that archived **Dr. Schwarzkopf** is run by Jason Leigh at the University of Illinois, Chicago, and is described in a *U.S. News & World Report* article from May 20, 2009: http://www.usnews.com/science/articles/2009/05/20/the-next-best-thing-to-you.html (accessed September 24, 2010). In addition, we interviewed the investigators via phone to learn more about the project.

151 In 2004, Dan Farber won an award from the American Society of Journalists and Authors for his article "The Man Who Mistook His Girlfriend for a Robot," which details David Hanson's work with **animatronic robots**. The story appeared in *Popular Science* on August 4, 2003, 60–70.

152 Clifford Nass's **multitasking** work is described here: Ophir, E., Nass, C. I., and Wagner, A. D., "Cognitive Control in Media Multitaskers," *Proceedings of the National Science Academy, 106* (2009): 15583–15587.

CHAPTER 10. DIGITAL FOOTPRINTS

PAGE

155 "**Hidden cameras**" in virtual reality do not require a camera-type device. Because the system renders—that is, redraws—the scene from scratch every frame, it must have information about the position and orientation of every object, including agents and avatars. Consequently, it is possible to render the scene from any viewpoint among those objects, given that all of the necessary spatial information is present.

156 Like all of the research reported in this book, the university Institutional Review Board, which determines whether experiments are ethical, reviewed and approved the ***Second Life* study**. Data from this study is reported in: Yee, N., Harris, H., Jabon, M., Bailenson, J. N. (in press),

"The Expression of Personality in Virtual Worlds," *Social Psychology and Personality Science.*

159 Albert "Skip" **Rizzo's ADHD work** is reported by: Parsons, T. D., Bowerly, T., Buckwalter, J. G., et al, "A Controlled Clinical Comparison of Attention Performance in Children with ADHD in a Virtual Reality Classroom Compared to Standard Neuropsychological Methods," *Child Neuropsychology, 13* (2007), 363–381.

160 The project on **online video game users** run by Dmitri Williams is funded by the National Science Foundation in the amount of $2,764,803: The Virtual World Observatory: Identifying Real World (RW) Characteristics from Virtual Behavior (2009–2012), Intelligence Advanced Research Projects Activity. PI: Dmitri Williams. Co-PIs: Noshir Contractor, M. Scott Poole, Jaideep Srivastava. The experimental data **predicting achievers** is here: Williams, D., Poole, M. S., and Contractor, N., "The Social Behaviors of Experts in Massive Multiplayer Online Role-playing Games," *Proceedings of the 2009 IEEE, SocialCom*, 326–331.

161 Paul Ekman's **Facial Action Coding System** is described in: Ekman, P., Friesen, W. V. & Hager, J. C. *The Facial Action Coding System* (Salt Lake City: Research Nexus, 2002).

163 The **facial tracking study of drivers** is here: Jabon, M. E., Bailenson, J. N., Pontikakis, E. D., Takayama, L. & Nass, C. (in press), "Facial Expression Analysis for Predicting Unsafe Driving Behavior," *IEEE Pervasive Computing.* The **facial tracking study of errors** is here: Jabon, M. E., Ahn, S. J. & Bailenson, J. N. (in press), "Predicting Performance on a Repetitive Task Through Automatic Analysis of Facial Feature Movements," *IEEE Journal of Intelligent Systems.*

164 Descriptions of Alex "Sandy" Pentland's work on **reality mining** can be found in: Pentland, A. *Honest Signals: How They Shape Our World* (Cambridge, Massachusetts: M.I.T. Press, 2008).

165 **Thin slicing** refers to using a small sample of data to predict behavioral outcomes. A description of the concept appears in: Ambady, N. & Gray, H. M. "On Being Sad and Mistaken: Mood Effects on the Accuracy of Thin-slice Judgments," *Journal of Personality & Social Psychology, 83* (4) (2002), 947–961.

165 James Pennebaker's study on **linguistic markers of deception** appears in: Newman, M. L., Pennebaker, J. W., Berry, D. S. & Richards, J. M., "Lying Words: Predicting Deception from Linguistic Styles," *Personality and Social Psychology Bulletin, 29* (2003), 665–675. His study on **linguistic markers of gender** is here: Newman, M. L., Groom, C. J., Handelman, L. D. & Pennebaker, J. W., "Gender Differences in Language Use: An Analysis of 14,000 Text Samples," *Discourse Processes, 45* (2003), 211–246.

166 For a description of the general methodology of using **language to predict behavior**, see: Tausczik, Y. R. & Pennebaker, J. W., "The Psychological Meaning of Words: LIWC and Computerized Text Analysis Methods," *Journal of Language and Social Psychology, 29* (2003), 24–54.

166 The study on using **language to predict military status** is described here:

Hancock, J. T., Beaver, D. I., Chung, C. K., Frazee, J., Pennebaker, J. W., Graesser, A. & Cai, Z., "Social Language Processing: A Framework for Analyzing the Communication of Terrorists and Authoritarian Regimes," *Behavioral Sciences of Terrorism and Political Aggression, 2* (2010), 108–132.

167 The study on using **language to predict psychopathy** can be found at: Woodworth, M., Hancock, J. T. & Porter, S., "Eat Your Words: Linguistic Analysis of Homicidal Psychopaths," presented at the 30th Congress on Law and Mental Health, Padua, Italy.

CHAPTER 11. THE VIRTUAL "JONES"

PAGE

170 For more information regarding **online volunteering,** read: Amichai-Hamburger, Y., "Potential and Promise of Online Volunteering," *Computers in Human Behavior, 24* (2008), 544–562.

171 Worried about being a **gadget?** Read: Lanier, J. *You Are Not a Gadget* (New York: Alfred A. Knopf, 2010).

171 The original study on **degrees of separation** is reported here: Travers, J. & Milgram, S., "An Experimental Study of the Small World Problem," *Sociometry, 32* (1969), 425–443.

174 Microsoft's **small-world** study was reported in the *Washington Post* by Peter Whoriskey, Saturday, August 2, 2008. See: http://www.washingtonpost.com/wp-dyn/content/article/2008/08/01/AR2008080103718.html.

175 For some **social-networking Web site data** see: http://hubpages.com/hub/Most-visited-Websites-in-the-World-Most-Popular-Websites-in-the-Internet and http://www.ebizmba.com/articles/social-networking-websites (accessed September 24, 2010).

175 **Statistics on the Internet** use are at http://www.internetworldstats.com/stats.htm (accessed September 24, 2010).

175 **Rural/urban** data from the U.S. Census Bureau.

176 **Number of Web sites data from:** http://royal.pingdom.com/2008/04/04/how-we-got-from-1-to-162-million-websites-on-the-internet/ (accessed September 25, 2010).

178 For arguments about the **similar brain** mechanisms, read: Grant, J. E., Brewer, J. A. & Potenza, M. N., "The Neurobiology of Substance and Behavioral Addictions," *CNS Spectrums, 11* (2006), 924–930.

179 Percentages **of Internet addiction problems** can be found in: Abou-jaoude, E., Lorrin, M. A., Gamel, N., Large, M. D. & Serpe, R. T., "Potential Markers for Problematic Internet Use: A Telephone Survey of 2,513 Adults," *CNS Spectrums, 11* (2006), 750–755.

179 Examinations of **social interaction data** can be found in: McKenna, Y. A. & Bargh, J. A., "Plan 9 from Cyberspace: The Implications of the Internet for Personality and Social Psychology," *Personality and Social Psychology Review, 4* (2000), 57–95. Its social impact is discussed in: Kraut, R., Kiesler, S., Mukhopadhyay, T., Scherlis, W. & Patterson, M., "Social Impact of the Internet: What Does It Mean?," *Communications of the ACM, 41* (1998), 21–22.

180 The ***FarmVille* infant murder plea** was reported by Catharine Smith in the *Huffington Post* on October 28, 2010. See: http://www.huffingtonpost.com/2010/10/28/Alexondra-v-tobias-farmville_n_775264.html (accessed November 13, 2010).

180 This report on **child neglect** was released by the Associated Press, July 15, 2007.

186 Did you or do you **gamble in college?** See if you agree with: LaBrie, R., Shaffer, H., LaPlante, D., and Wechslet, H., "Correlates of College Student Gambling in the United States," *Journal of American College Health, 52* (2003), 53-62, and Looney, E., "The Surge in Online Gambling on College Campuses," *New Directions for Student Services* (2006), 53–61.

186 Completely rational people may wonder **why people gamble** in casinos, where the establishment has the advantage. Gambling motivation is discussed by: Raichlin, H., "Why Do People Gamble and Keep Gambling Despite Heavy Losses?," *Psychological Science, 1* (1990), 294–297.

188 The debate over **media violence** is long-lived. See, e.g., **Bushman**, B. & Gibson, B., "Violent Video Game Causes and Increase in Aggression Long After the Game Has Been Turned Off," *Social Psychology and Personality Science, 1* (2010), 168–174, and **Freedman**, J. L. *Media Violence and Its Effect on Aggression: Assessing the Scientific Evidence* (Toronto: University of Toronto Press, 2002).

CHAPTER 12. VIRTUALLY USEFUL

PAGE

191 The book on how **virtual worlds using avatars will change business and other institutions** is: Reeves, B. & Read, L. J. *Total Engagement* (Cambridge, Massachusetts: Harvard Business Press, 2009).

192 The *Time* magazine article listing the **fifty worst cars** is here: http://www.time.com/time/specials/2007/0,28757,1658545,00.html (accessed September 25, 2010).

193 The keynote speech at IEEE Virtual Reality by **Elizabeth Baron** was called "Successes and Challenges on Using VR in Product Design and Engineering," delivered on March 17, 2009. It is available for download here: http://vgtc.org/wpmu/vr09/2009/03/20/keynote-speaker-elizabeth-baron/ (accessed September 25, 2010).

195 On **Super Bowl viewing,** see an article by Holly Sanders Ware in the *New York Post*, January 20, 2010, titled "More People Watch Super Bowl for Ads Than Game."

195 **Volkswagen's** 1960 advertising budget data were retrieved from: McLeod, Kate. *Beetlemania: The Story of the Car That Captured the Hearts of Millions* (New York: Smithmark Publishers, 1999). The 2005 data was retrieved from a *New York Times* article: http://www.nytimes.com/2005/09/07/business/media/07adco.html (accessed September 25, 2010).

196 The ideas presented on **virtual reality and marketing** were contributed by Bruce Miller, CEO of Market Data Corporation, Chicago, Illinois.

198 On December 1, 1997, *Industry Week* reported on **Kimberly-Clark's use of virtual reality** for market research: http://www.industryweek.com/articles/Kimberly-clark_embraces_virtual_reality_15349.aspx (accessed September 25, 2010).

200 For a detailed description of how basic economic **principles operate in *Second Life*,** see: Castronova, E. *Synthetic Worlds: The Business and Culture of Online Games* (Chicago: University of Chicago Press, 2005).

200 For a more detailed discussion of **product placement in video games**, see this article published by MSNBC on July 21, 2006: http://www.msnbc.msn.com/id/13960083/ (accessed September 25, 2010).

200 The **courtroom applications of virtual reality** are described in detail here: Bailenson, J. N., Blascovich, J., Beall, A. C. & Noveck, B., "Courtroom Applications of Virtual Environments, Immersive Virtual Environments, and Collaborative Virtual Environments," *Law and Policy, 28* (2) (2006), 249–270.

202 Most **lineups in the United States** don't use live actors, because it is too difficult and expensive to find foils that match the physical description of the suspect, as pointed out by: Wogalter, M. S., Malpass, R. S. & McQuiston, D. E., "A National Survey of U.S. Police on Preparation and Conduct of Identification Lineups," *Psychology, Crime & Law, 10* (1) (2004), 69–82.

204 Our study varying **contextual cues in virtual reality** is reported here: Bailenson, J. N., Davies, A., Beall, A. C., Blascovich, J., Guadagno, R. E. & McCall, C., "The Effects of Witness Viewpoint Distance, Angle, and Choice on Eyewitness Accuracy in Police Lineups Conducted in Immersive Virtual Environments," *PRESENCE: Teleoperators and Virtual Environments, 17* (3) (2008), 242–255.

204 The statistics pertinent to eyewitness testimony and **wrongful conviction** can be found here: *The Innocence Project—Understand the Causes: Eyewitness Misidentification,* http://www.innocenceproject.org/understand/Eyewitness-Misidentification.php (accessed September 25, 2010).

205 The study on **surgical training** is reported in: O'Toole, R., Playter, R., Blank, W., Cornelius, N., Roberts, W., and Raibert, M., "A Novel Virtual Reality Surgical Trainer with Force Feedback: Surgeon Vs Medical Student Performance," in Salisbury, J. K., and Srinivasan, M. A. (eds.). *Proceedings of the Second Phantom Users Group Workshop* (Delham, Massachusetts: M.I.T. Press, 1997), 73–75.

206 The **physical therapy** application of virtual reality is described by: Deutsch, J., Latonio, J., Burdea, G., and Boian, R., "Post-Stroke Rehabilitation with the Rutgers Ankle System—A Case Study," *Presence, 10* (4) (2001), 416–430.

207 A description of **Admiral Nelson's phantom limb** syndrome appears here: Goody W., "Admiral Lord Nelson's Neurological Illnesses," *Proceedings of the Royal Society of Medicine, 63* (1970), 299.

207 Hunter Hoffman's **pain research in virtual reality** can be found at: Hoffman, H. G., "Virtual-reality Therapy," *Scientific American, 291* (2) (2004), 58–65.

208 Albert "Skip" Rizzo's work on **Virtual Iraq** is detailed in: "Virtual Iraq," *New Yorker,* May 19, 2008, 32–37.

211 A description of the technological history of the **Link flight simulator** is here: Jaspers, Henrik, "Paper to Royal Aeronautical Society Conference," available online at http://home.wanadoo.nl/hjaspers000 (accessed September. 23, 2010).

213 The *New York Times* discusses the success of the video game ***America's Army***: http://www.nytimes-se.com/2009/07/04/recruiting-tool-cancelled/ (accessed September 25, 2010).

214 On August 6, 2008, *Wired* described a **protest against *America's Army***: http://www.wired.com/gamelife/2008/08/ubisoft-protest/comment-page-3/ (accessed September 25, 2010).

215 Mike Zyda talks about how to use ***America's Army* for recruiting specific skill sets.** See Zyda, M., "The Naval Postgraduate School MOVES Program—Entertainment Research Directions," *Proceedings of Summer Computer Simulation Conference*, Vancouver, BC, Canada, July 2000, 1–6.

215 The *New York Times* describes the Arabic video game ***Special Force***: http://www.nytimes.com/2003/05/18/international/middleeast/18VIDE.html (accessed September 25, 2010). An example of using virtual reality to **teach soldiers about culture** is: Lane H. C., Hays, M., Core, M., Gomboc, D., Forbell, E., Auerbach, D., and Rosenberg, M., "Coaching Intercultural Communication in a Serious Game," International Conference on Computers in Education, Putrajaya, Malaysia, 2008.

217 The *New York Times* discusses the military simulation ***Urban Resolve***: http://www.nytimes.com/2003/05/18/international/middleeast/18VIDE.html (accessed September 25, 2010).

218 For a review of the work on **phobia treatment in virtual reality,** see: Riva, G., Wiederhold, B. K., and Molinari, E. *Virtual Environments in Clinical Psychology and Neuroscience: Methods and Techniques in Advanced Patient-therapist Interaction* (Amsterdam: IOS Press, 1998).

220 The study on **public-speaking anxiety** in virtual reality is reported here: Slater, Mel, Pertaub, David-Paul, Barker, Chris, Clark, David M., "An Experimental Study on Fear of Public Speaking Using a Virtual Environment," *Cyberpsychology & Behavior, 9* (5) (2006), 627–633.

221 For a discussion of the link between **virtual reality research and science fiction,** see: Bailenson, J. N., Yee, N., Kim, A. & Tecarro, J., "Sciencepunk: The Influence of Informed Science Fiction on Virtual Reality Research," in Grebowicz, Margret (ed.). *The Joy of SF: Essays in Science and Technology Studies* (Chicago: Open Court Publishing, 2007), 147–164.

222 For a description of the use of **virtual reality in archaeology,** see: Sanders, D. H., "Virtual Worlds for Archaeological Research and Education," in Dingwall, L., Exon, S., Gaffney, V., Laflin, S., Van Leusen, M. (eds.), *Archaeology in the Age of the Internet* (Oxford, England: British Archaeological Reports, Int. Series, S750, 1999), 265.

224 **Letter from PETA** can be retrieved from: http://www.peta.org/b/thepetafiles/archive/2008/08/15/a-peta-theme-park.aspx (accessed September 25, 2010).

225 The study on the **virtual window** is reported in: Ijsselsteijn, W., Oosting,

W., Vogels, I., de Kort, Y., van Loenen, E., "Look at or Looking Out: Exploring Monocular Cues to Create a See-through Experience with a Virtual Window," *Proceedings of the 2006 Presence Conference*, Cleveland, Ohio.

CHAPTER 13. VIRTUAL YIN AND YANG

PAGE

229 Some of the very early and surprising research on **sign language and a chimpanzee** is reported in: Gardner, R. A. & Gardner, B. T., "Teaching Sign Language to a Chimpanzee," *Science, 165* (1969), 664–672. Likewise, a project using a gorilla is reported in: Patterson, F. G., "The Gestures of a Gorilla: Language Acquisition in Another Pongid," *Brain and Language, 5* (1978), 72.

230 **Trippi,** J. *The Revolution Will Not Be Televised: Democracy, the Internet and the Overthrow of Everything* (New York: HarperCollins, 2004).

231 Estimate for **identity theft** from: Kahn, C. M. & Roberts, W., "Credit and Identity Theft," *Journal of Monetary Economics, 55* (2008), 251–264.

232 The **Drew indictment was** reported on May 16, 2008, in the *New York Times.* See: Steinhaueur, Jennifer, "Woman Indicted in MySpace Suicide Case," http://www.nytimes.com/2008/05/16/US/16myspace.html.

233 For a discussion of the relationship between **scientists and science fiction,** read: Bailenson, J. N., Yee, N., Kim, A. & Tecarro, J., "Sciencepunk: The Influence of Informed Science Fiction on Virtual Reality Research," in Grebowicz, Margret (ed.). *The Joy of SF: Essays in Science and Technology Studies* (Chicago: Open Court Publishing, 2007), 147–164.

235 Kane, H., McCall, C., Collins, N. & Blascovich, J., "Understanding the Effects of **Social Support** from a Significant Other in an Immersive Virtual Environment," paper presented at the annual conference of the Society for Personality and Social Psychology, February 2009, Tampa, Florida.

238 A classic paper about **personal space** is: Mehrabian, A., "Personal Space," *Journal of Experimental Social Psychology, 1* (1965), 237.

241 Morgan, R. ***Altered Carbon*** (New York: Random House, 2002).

241 **Pope quote** in the Montreal *Gazette* on October 8, 2010. See: http://www.montrealgazette.com/news/technologies+confuse+reality+fiction+Pope/3643380/story.html (accessed November 13, 2010).

242 The **numbers** came up in Google searches on November 12, 2010

242 **Bob Hyatt and Douglas Estes** didn't see eye-to-eye on the value of churches in virtual reality. Their points of view are more fully represented in: Estes, D. *SimChurch: Being the Church in the Virtual World* (Grand Rapids, Michigan: Zondervan, 2009), and Bob Hyatt's position as stated in http://www.outofur.com/archives/2009/08/there_is_no_vir.html (accessed September 24, 2010).

245 Stephenson, N. ***The Diamond Age*** (New York: Bantam Dell, 1995).

246 **Dede,** C., "Immersive Interfaces for Engagement and Learning," *Science, 323* (2009), 66.

249 For more on problem-solving and **patterns of eye movements,** read Grant,

E. R. & Spivey, M. J., "Eye Movements and Problem Solving," *Psychological Science, 14* (2003), 462.

251 Reynolds, G. ***An Army of Davids:*** *How Markets and Technology Empower Ordinary People to Beat Big Media, Big Government, and Other Goliaths* (Nashville, Tennessee: Thomas Nelson, 2006).

251 The story, **Political Factors Complicate China's Clout in Mideast**, NPR Story, April 4, 2008. http://www.npr.org/templates/story/story.php?storyId=89385642&ps=rs (accessed September 20, 2010).

252 For more depth on the **Zapatista rebellion,** read: Morello, H. J., "E-(re)volution: Zapatistas and the Emancipatory Internet," *Contra corriente: A Journal on Social History and Literature in South America, 4* (2007), 54–76.

CHAPTER 14. MORE HUMAN THAN HUMAN

PAGE

257 Damásio, A. ***Descartes Error:*** *Emotion, Reason, and the Human Brain* (New York: Quill, 1995).

258 For more information, read: **Dennett, D. C**. *Brainstorms: Philosophical Essays on Mind and Psychology* (Cambridge, Massachusetts: Bradford Books, 1978).

258 **Nozick**, R. *Invariance: The Structure of the Objective World* (Cambridge, Massachusetts: Harvard University Press, 2001).

259 For the bible of psychological and psychiatric disorders, see: American Psychiatric Association, Task Force on DSM-IV, *Diagnostic and Statistical Manual of Mental Disorders*, 4th ed. (2000). Note that **multiple personality disorder** is now labeled "Dissociative Identity Disorder."

259 Schreiber, F. R. ***Sybil*** (New York: Warner Paperback Library, 1974).

260 For more on the **normality of multiple personalities in virtual reality:** Turkle, S. *Life on the Screen: Identity in the Age of the Internet* (New York: Simon & Schuster, 1995).

260 See Moor, J. H. & Bynum, T. W. ***Cyberphilosophy****: The Intersection of Philosophy and Computing* (Oxford, England: Blackwell, 2002).

260 See Lastowka, G. ***Virtual Justice****: The New Laws of Online Worlds* (New Haven, Connecticut: Yale University Press, 2010).

261 Clark, A. C. ***The City and the Stars*** (New York: Harcourt Brace Jovanovich, 1953).

261 **"The Year 2100"**: http://mysite.verizon.net/wsbainbridge/system/software.htm (accessed September 24, 2010).

261 Another example of people **archiving their lives digitally** is the Shoah Project, see http://college.usc.edu/vhi/otv/otv.php (accessed September 24, 2010).

INDEX

About the authors

About the book

Read on

Insights, Interviews & More . . .

Meet Jim Blascovich

Jim Hammock

JIM BLASCOVICH is a distinguished professor of psychological and brain sciences and codirector of the Research Center for Virtual Environments and Behavior, which he cofounded with Jack Loomis, a perceptual scientist, at the University of California–Santa Barbara in 1997. He held academic positions at the University of Nevada, Marquette University, and SUNY Buffalo before coming to UCSB in 1995. From 2010 to 2011 he was a visiting scholar at the Center for Advanced Study in the Behavioral Sciences at Stanford. Jim is a past president of both the Society for Personality and Social Psychology and the Society of Experimental Social Psychology. He is a member of the Academy of Behavioral Medicine Research, a charter fellow of the American Psychological Society, and a fellow of the American Psychological Association. He is a recipient of the Gordon Allport Intergroup Relations Prize, the Inaugural Australasian Teaching Fellowship, and an Erskine Fellowship at the University of Canterbury in Christchurch, New Zealand. He has served on several

National Research Council panels and numerous editorial boards. His research has been continuously funded by the National Science Foundation for more than twenty years and has been supported by the National Institutes of Health, the Army Research Laboratory, and other agencies. He has more than 140 publications, including four books.

His major research interests include social influence within technologically mediated environments and social neuroscience. Guided by his theoretical model of social influence within immersive virtual environments, Jim investigates social influence processes in virtual reality, including conformity, nonverbal communication, collaborative decision making, and leadership. He has also developed the biopsychosocial model of challenge and threat, guiding research on social psychological processes within and outside of virtual environments.

Meet Jeremy Bailenson

Debbie Hill

Jeremy Bailenson is the founding director of Stanford University's Virtual Human Interaction Lab and an associate professor in the Department of Communication at Stanford. He earned a BA cum laude from the University of Michigan in 1994 and a PhD in cognitive psychology from Northwestern University in 1999. After receiving his doctorate, he spent four years at the Research Center for Virtual Environments and Behavior at the University of California–Santa Barbara as a postdoctoral fellow and then as an assistant research professor.

Bailenson's main area of interest is the phenomenon of digital human representation, especially in the context of immersive virtual reality. He explores the manner in which people are able to represent themselves when the physical constraints of body and veridically rendered behaviors are removed. He also designs and studies collaborative virtual reality systems that allow physically remote individuals to meet in virtual space, and explores the manner in which these systems change the nature of verbal and nonverbal interaction.

His findings have been published in more than seventy academic papers in the fields of communication, computer science, education, law, political science, and psychology. His work has been consistently funded by the National Science Foundation for more than a decade, and he also receives grants from various Silicon Valley and international corporations. Bailenson consults regularly for government agencies, including the Army, the Department of Defense, the National Research Council, and the National Institutes of Health, on policy issues surrounding virtual reality.

Six Commandments for Virtual Life

THERE IS LITTLE DOUBT the digital virtual revolution has begun in earnest. Technology innovations become outdated almost as quickly as they appear. Even people like us, who study virtual technology for a living, have a hard time keeping up. It is going to be one wild and bumpy ride filled with highs and lows, or as we like to call them, the yin and yang. Let's consider both here. The latest and greatest virtual technologies can give us new toys to play with and revolutionary tools to work with. However, they come at a cost, for these gadgets can lead us into situations that cause embarrassment, indictment, and even addiction. Below, we offer six rules that readers can follow to smooth the virtual roller-coaster ride.

“Even people like us, who study virtual technology for a living, have a hard time keeping up.”

The Yins

1. Make Virtual Reality Work For You.

The number of practical applications virtual reality offers is growing by leaps and bounds. Virtual therapies can treat fears of flying, public speaking, heights, and spiders. Virtual shopping applications permit shoppers to try before buying, seeing how designer eyeglass frames look on their faces, how clothes fit, or even how they will look sitting in a brand-new sports car. Immersive video games can train one to play a wide range of sports, from Ping-Pong to cricket. Don't allow fear of virtual reality to ruin the fun and advantages of all the amazing applications out there. They are not just for geeks, nerds, degenerates, or lazy people!

2. Consider Your Virtual Legacy.

The practice of creating photo albums and producing home videos is so entrenched in society that just about everyone archives these two-dimensional versions of themselves for future generations to experience. But why not preserve yourself in a more thorough, immersive manner? Three-dimensional scanners, similar to the ones used to immortalize Brad Pitt's character in *The Curious Case of Benjamin Button* as well as actors in new video games such as *LA Noir*, are relatively common and inexpensive. Commercial centers exist where you can scan your head and body; you can then take home the model of your digital body on a thumb drive. Today's typical grandparent believes that the idea of cryogenics—freezing their brains for future reincarnation—is a bit much. On the other hand, virtual reality makes reincarnation via 3-D avatars more acceptable—the perfect compromise for virtual immortality without going over the top.

3. Mix the Physical and the Digital.

It's a good idea to make sure that your virtual life stays connected to the physical ones of the people who matter most to you. Video games, perhaps the most popular use of virtual reality today, are consumed more often by children than television programming and movies combined. Parents who play video games *with* their children can share some of their children's favorite activities. In the same way that families watch sports and movies on television together, making virtual experiences a group activity is a healthy way to enjoy the medium. The truth is that there is a young but maturing generation who look for virtual spaces for entertainment before they look for physical ones. Enjoying those virtual experiences is a way to connect positively with family and friends.

The Yangs

4. Watch Your Digital Footprint.

Some time ago, an undergraduate was applying to a PhD program. During the admissions committee meeting in which professors consider which of the many applicants should be interviewed for possible admission (typically less than 5 percent get interviewed), a professor decided to Google the applicant. He found an interesting entry on the applicant's blog. The student wrote that he was only applying to this specific graduate school so he could say he got in—he indicated that he had no intention of accepting the school's offer. It's easy to see what happened next—the student got rejected, though he later indicated having no memory of writing the damning message on his blog and maintained he would have loved to attend that particular school.

If text can be incriminating, imagine what kind of drama photographs can stir up. A professor we know conducts an exercise every year that demonstrates effectively the possibility of calamity via ▶

Six Commandments for Virtual Life ***(continued)***

social networking. His teaching assistants search his students' Facebook profiles—the ones that are public and can be seen without special permission. The assistants consistently find many pictures that would mortify students if they thought their professors could see them. The assistants digitally blur out the students' faces so they are unrecognizable to the professor or fellow classmates. On the first day of class, the professor scrolls through a few pictures—some of students doing illegal drugs, others partially nude—and tells the class, "These are pictures of some of you in this class. The teaching assistants found them by searching for your name and it took less than an hour." The professor makes sure he doesn't know who the students are, as he doesn't want to be biased in his grading. Nonetheless, every year, these students are chagrined, but hopefully they learn a lesson. And it is not just professors who are curious enough to do such searches. Think about significant others, parents, children, loved ones, and employers, as well.

The most misunderstood aspect of virtual experience is anonymity. Consider the plight of former New York Congressman Anthony Weiner. He used Twitter for sexting; that is, sending digital erotic messages and explicit images of himself to women around the country. Like most people, he undoubtedly assumed that Twitter activities are ephemeral and anonymous, but nothing could be further from the truth.

As we highlight in the "Digital Footprints" chapter, all virtual activity can be tracked and used to identify who you are and what you have done. Of course, names and photos can be controlled and even faked or borrowed by users. However, as virtual experiences become more immersive, it will become even more difficult to control which aspects of one's identity become public. For example, video games can track every move players make and record every word players say. Games can even recognize an individual's typing and controller styles and habits. Newer video game platforms actually involve a camera that captures players' behaviors. More and more of work, play, and social life occurs in virtual spaces. An understanding and awareness of how many secrets we are leaving on the table is critical.

The good news is that digital footprints can be managed. Don't use realistic, high-resolution photographs on websites, in social networking profiles, or to create an avatar. Using just a few pictures of a person, it is possible for someone else to create a 3-D photorealistic model. Once made, the possibilities for manipulation are many—ranging from creating compromising poses and actions (think what an angry ex-boyfriend or -girlfriend could do with such technology) to sending out tailored political advertisements or online used-car showrooms that imperceptibly blend a voter's face into the candidate's or salesman's face.

The solution is simple—"thumbnail" your images by reducing their resolution, and remind yourself constantly that everything you do online can be tracked by someone or some machine. Everything. So even if you're in a coffeeshop on a rented computer using a random username in a virtual chatroom using fairly advanced encryption software, be careful what and how you type. The virtual world offers the possibility of experiencing things that are not possible in the physical world, and while they are tempting and possibly rewarding, anonymity is never guaranteed.

5. Be Wary of Addiction.

The number of hours that people engage in virtual behavior every day is staggering. In the United States, children between the ages of eight and eighteen spend, on average, more than eight hours per day using digital media outside the classroom. In some countries, it's even higher. Think about that—more waking hours are spent using digital media than not. As the online world becomes more perceptually immersive—that is, more engrossing and rewarding—online addictions become more pervasive.

Simply thinking about virtual experiences as a potentially addictive behavior is a great place to start. Internet pornography can become so addictive that people prefer the virtual images—even though they are only two-dimensional movies—to having sex with real people. Online gambling is similarly addictive. People can get addicted to innocent games such as Farmville. As these experiences become even more multisensory, and people can touch, hear, smell, and taste them, will users still be able to relate to actual people?

Another good practice is to simply monitor the amount of time one spends in virtual places. Between e-mails, texts, cell phones, social networking, etc., the hours add up quickly. The total time one comes up with is often disheartening because it is so high. Simply being aware of the amount of time spent in virtual activities can be half the battle.

Go cold turkey for an extended period of time. During recent trips to New Zealand and rafting the Colorado River through the Grand Canyon, respectively, each of us decided to go "without" for relatively extended periods, that is, without touching or listening to a digital media device—no e-mail, cell phone, MP3 player, World Wide Web, etc. It took three full days for the grasping motion—the involuntary twitching reach of the hand into the pocket to check the smartphone—to subside. The need to maintain the virtual connection was so entrenched it had burrowed its way into motor memory. ▶

Six Commandments for Virtual Life *(continued)*

6. Look for Yourself!

In Chapter 8, we described how advertisements, especially political ones, become more effective when their messages are tailored specifically to targeted individuals. In 2004, we sent virtual campaign advertisements to individual voters in which George Bush's or John Kerry's face was morphed to look like that individual voter, but only enough to register subliminally. Some people saw versions of Kerry that absorbed about a third of their facial features using a morphing technique, and others saw versions of Bush that looked similarly like them. Even though the voters did not consciously recognize their own faces in the morph, they were more likely to vote for the candidate that looked like them. Similarly, if a virtual character mimics your gestures—for example, by analyzing your movements while you use a video game platform such as the Microsoft Kinect—you will like that character more, even if you aren't conscious of the mimicry. In Chapter 8, we described a series of studies in which a consumer's face is taken from a public database and placed in an advertisement, such that the consumer becomes an unwitting endorser of the product. This is not just an academic research strategy—since reading our published work, the Silicon Valley company LinkedIn has incorporated this feature into some of its advertising. A job candidate is now lured toward certain job descriptions by seeing her face displayed within the description. In this way LinkedIn increases the emotional connection between the consumer and the job description, and scores more applications.

So if a character in an online advertisement, dating profile, or virtual world seems particularly appealing for a reason you just can't put your finger on, it may be time to look a bit closer at the face or gestures of that character. Understanding the consequences of seeing your own face in advertisements is crucial to being able to resist the persuasiveness of the message.

Our research has demonstrated that knowledge is power. If a person being manipulated explicitly notices the strategy, that is, becomes aware that the candidate looks more like him, the manipulation backfires. Simply by detecting these manipulations, we can protect ourselves.

Think of the advice above as a digital seatbelt and use it to virtually travel sanely and safely. But be warned that these suggestions are not guaranteed to protect users, nor are they exhaustive. They just increase the likelihood of positive virtual experiences and decrease the chances of bad ones.

How Did You Become a Virtual Reality Pioneer?

Jim Blascovich

Life puts us on unplanned as well as planned paths—more of the former, in my case. A freshman math major and sorry monolingual to boot, stuck with an essentially monolingual East Asian calculus instructor, I switched to psychology.

I don't know what I would have done with only a BS in psychology, but I never had the chance to find out, as the Vietnam war, the draft, and a good mentor pushed me to apply to graduate school. Rejection after rejection arrived until finally, with only one possibility left, I sent a telegram (the instant message of its time) to the University of Nevada giving them my postgraduation address and directing Western Union to send me a copy of it, too. The copy arrived the next day, but I left it unopened. A few days later, I actually opened it to find not a copy but an acceptance and fellowship offer. Joyful beyond belief, I accepted via Western Union without delay.

Graduate school was a wonderful and enlightening experience. I was hooked quickly. Yet being without a draft deferment, I had to consider my options. Canada? Nah! Flunk my preinduction physical? Nah! ROTC? Maybe! But, serendipitously, I heard a radio advertisement for enlistment in the Nevada Air National Guard while driving home one day. This was late 1968. To get into the Guard elsewhere, one had to be connected, but in Reno it was open and fair. I did a U-turn, drove to the base, and signed up (actually the same week as "W," though I attended all required meetings). I left for ▸

six months of basic and specialized training, became a "weekend warrior," and then returned to grad school.

Being a graduate student in the late sixties and early seventies was an often raucous experience. "Hell no! We won't go!" rang in my ears and in my heart. But the Air Guard was a heady experience as well. I had to manage conflicting identities without the benefit of the Internet or digital VR. Graduate school was my yin to the Air Force's yang, and vice versa.

In grad school I became bilingual, then trilingual, and even quadrilingual, learning BASIC, SNOBOL, and FORTRAN programming. "Flying a typewriter" gave me the ability to enter data onto Hollerith cards (look it up) faster than anyone I knew. Programming allowed me to analyze data and, with SNOBOL, even language. At the dawn of the solid-state electronics age, I was also drawn to digital gadgets that I invented but did not patent, including a device that when eventually thought up by someone else made him a zillionaire.

After graduate school, I earned tenure at Marquette University in fast order. I was quite proud of myself, but was caught off guard by one of my assistant professor colleagues, who snipped, "Congratulations, Blascovich! You've got a life sentence—the rest of us are only doing one to six." Snide as it was, the statement had merit. Realizing that tenure should not be a life sentence, I vowed not to let it keep me from exploring new horizons.

Soon I was on sabbatical, "temporarily" moving to SUNY Buffalo. There I fell in love, never to return to Marquette. I wandered in the desert of academe for a few years, taking so-called soft-money jobs in the medical school and the school of allied health (earning tenure #2) and returning to my interest in all things digital by starting a computer consulting and software business at the beginning of the microcomputer revolution. I proved to myself that I could make money selling my expertise. But I also realized it wasn't the secret of life. Fortunately, a few academic doors opened and I walked happily through them, joining the psychology faculty at Buffalo (tenure #3) and establishing myself as a social psychophysiologist.

In 1992 I was again on sabbatical, this time at the University of California–Santa Barbara. It was as good a place to be, in all respects, as one might imagine. At the end of the visit, while sitting in one of the many outdoor cafés, I wondered aloud, "How come all these people get to live here?" To make a long story short, I joined the faculty at UCSB three years later (tenure #4).

Early on, I experienced another open door—this time a literal one, to my colleague Jack Loomis's perception laboratory. Inside, a young graduate student wore an interesting contraption involving a head-mounted display that totally covered her eyes and a backpack with short poles capped by low-voltage lights. She was walking around, turning here and there, sometimes moving forward or reversing course. Clearly she was engaged in something that she was seeing but that I wasn't.

Computers, gadgets, and serendipity immediately colluded to grab my mind. What was going on? What was she doing? Where was she? I had to know. So I asked Jack if I could see. He happily agreed. I donned the equipment, finding myself on an immense flat prairie open to a blue-gray sky to the horizon. I could walk freely about and did as Jack instructed.

"OK, what's so special here, Jack?"

"Look down," he commanded.

I did, and what I saw changed my life. An open pit about three meters square appeared in the ground. As Jack coached me to its edge, I felt the same fright as if I were on the edge of a real abyss. Jack asked if I felt like stepping off the edge. I refused. He said he could turn off "gravity" in the virtual world so I could walk on "air" above the pit. No way. He put down a virtual plank that spanned the pit and asked me if I could walk across it. No way . . . well, ok, I'll try. To make a long story shorter, I eventually made it across the pit, sweating and extending my arms out for balance.

Upon reflection, I realized that my mind could not control my fear. No matter what I told myself, the pit was still frighteningly real. More important, I realized what a boon this technology could be for social psychological research, and began a discussion with Jack that led within a year to the founding of the interdisciplinary Research Center for Virtual Environments and Behavior at UCSB. The Center was originally located in the old psychology building, then in less tight quarters in it, and it eventually moved to its permanent home, a three thousand square foot facility in the new psychology building. The lab maintains cutting-edge equipment, tracking systems, and software and has benefited from funding from many federal agencies, especially the National Science Foundation.

In its fifteen years of existence, we've produced much research and trained numerous graduate students and postdoctoral fellows as well as researchers from around the world. A VR technology company, WorldViz, was spun off from ReCVEB some ten years ago (in which—true to form—I have no financial interest).

The secret to what success I've enjoyed lies especially with my closest colleagues: Jack Loomis, a true visionary who taught me a lot about perception, philosophy, and virtual reality; Andy Beall, whom we hired as a scientist and technology director—a sheer genius who truly is the mother of VR invention in my mind; and Jeremy Bailenson, a ▶

How Did You Become a Virtual Reality Pioneer? ***(continued)***

postdoc—whom the world knows from his days as a wunderkind of VR research at UCSB, an established scholar at Stanford, and coauthor of this book.

I had the good fortune of being in the right place at the right time and sometimes taking advantage of it, and, oh yes, knowing that tenure is not a life sentence.

Jeremy Bailenson

I arrived at Stanford in the fall of 2003, and was tasked with creating "THE" virtual reality lab at such a prestigious university. The task was daunting—the square footage I could work with was limited and my equipment budget was modest. But the real estate was prime—I was lucky enough to be given an enviable space that overlooked the front entrance of the gorgeous gate to the university, with Rodin sculptures spotting the courtyard, gazing thoughtfully toward the entrance to my building.

The lab grew slowly, being run in its first two years by a team of three great people—PhD student Nick Yee, who was the genius behind much of the work; a programmer named Jerry Yu, who built Python scripts; and a woman named Claire Carlson, who managed all the human resources aspects of the lab. This bare-bones team produced much of the work that provided a foundation for the lab.

As the years went by, the resources of the lab slowly grew. Science requires dollars, and they rolled in from a number of sources. Stanford provided enough money when I arrived to buy some head-mounted displays (each one costs as much as a new car) and the hardware to power them. Worldviz LLC, a company that creates VR software and hardware, gave me the educational discounts required to get started. Of course the National Science Foundation was pivotal, funding many of the ideas which at the time were outlandish to some people—for example, how spending time in avatars could affect your self-esteem in the physical world—but that by now have headlined just about every newspaper. The Silicon Valley industry was quick to give gifts for various projects, such as using avatars to create diversity training software. Stanford itself was consistently generous, funding me to mentor brilliant undergraduates who do the lion's share of the virtual content creations for the lab. PhD students sought out Stanford in order to work there. Looking back at the bare-bones team of three, I marvel at the number and talent of the scholars who now make up the lab.

I am sometimes asked, "When did you take over running the Virtual Human Interaction lab?" This always makes me smile, reinforcing the hope that I have created an institution that will last longer than I do.

Studying the psychological implications of virtual life is not yet a required class in school, but it may be someday soon.

In 2009, I received tenure (huge sigh of relief), and the powers that be in the dean's office decided to commit substantial resources toward a complete overhaul of the lab. The construction of this facility, which cost seven figures to build, was finally finished in July of 2011, and in my humble opinion it truly provides one of the most compelling combinations of three senses—sight, sound, and touch—of any virtual reality lab in the world. The lab has drawn an impressive list of VIP guest visitors, including CEOs, government delegations, actresses, military officers, writers, and "famous VR guys." It's fantastic to have a lab, finally, that is worthy of being dubbed "the VR Lab" at Stanford.

My Relationship with Neuromancer

Readers may take issue with the number of science fiction references in this book. One in particular stands out. My relationship with *Neuromancer* is a complicated one. On the one hand, it was without a doubt what inspired me to become a scholar of avatars. William Gibson's unprecedented vision of cyberspace redefines what it means to be human—mortality is optional, people can transform their gender, age, and identity at the drop of a hat, and the notions of pleasure and pain move beyond the flesh. Indeed, these themes are pervasive throughout my research and throughout *Infinite Reality.*

My reliance on this work of fiction written over two decades ago has been substantial. I first read it in high school, and like many Stanford students who are now forced to read it in my courses, I didn't really get it. It's a challenging read on the first try, and some of the big ideas take a few reads before they grab hold. I didn't pick the book up again for about a decade, in my fifth year of graduate school, when I was floundering without direction in research, running cognitive psychology experiments, and designing computer programs that mimicked thought. But before I dropped out of academia altogether, one job advertisement resonated with what I had read in *Neuromancer*, and I decided to give university life one more try, taking a research position at UCSB studying avatars.

Since then, *Neuromancer* has been my crutch. Large government grants have been awarded to me for building and testing Gibson's ideas. Academic papers are improved by Gibson quotes that sum up the big ideas of the research. PhD students walk out of my office with a copy when searching for dissertation topics. Undergraduates who can't imagine the world without the "cyberspace" Gibson predicted (or perhaps facilitated) grumble about my using it as a textbook in my lecture classes. ▸

How Did You Become a Virtual Reality Pioneer? *(continued)*

Dixie Flatline is a character from *Neuromancer* who has his avatar resurrected—after he passes away physically, his personality and consciousness are gathered from data he had left behind. Dixie was the inspiration for many studies run at Stanford and two chapters in *Infinite Reality* ("Digital Footprints" and "Virtual Immortality"). As Gibson predicts, and studies show, virtual immortality may not always be a good thing. Dixie Flatline sums it up better than I ever could when he declares: "I wanna be erased."

Without *Neuromancer*, the world of virtual reality as a whole, and certainly *Infinite Reality*, would look very different.